Biochemistry and Food Science

E.N. Ramsden

B.Sc., Ph.D., D.Phil.

Formerly of Wolfreton School, Hull

Stanley Thornes (Publishers) Ltd.

First published 1995 by
Stanley Thornes (Publishers) Ltd,
Ellenborough House,
Wellington Street,
CHELTENHAM,
GL50 1YD

A catalogue record for this book is available from the British Library.

ISBN 0 7487 1806 0

The front cover shows a computer graphics close-up of an ATP molecule bound to phosphoglycerate kinase, a glycolytic enzyme.

Typeset by Tech-Set, Gateshead, Tyne & Wear.
Printed and bound in Great Britain at Scotprint, Musselburgh

CONTENTS

PART 1: BIOCHEMISTRY

CHAPTER 1: CELLS

CHAPTER 2: PROTEINS

CHAPTER 3: CARBOHYDRATES

CHAPTER 4: LIPIDS

PART TWO: FOOD SCIENCE

CHAPTER 11: FOOD PROCESSING

CHAPTER 12: FOOD PRESERVATION

CHAPTER 13: ADDITIVES

CHAPTER 14: LEGISLATION

CHAPTER 15: HUNGER

PREFACE

Biochemistry and Food Science has been written to match the 1996 syllabuses for the following A-level modules:

Northern Examinations and Assessment Board:
  Module Ch 9: The chemistry of living systems and food
University of Cambridge Local Examinations Syndicate:
  Syllabus 9254: Biochemistry option
              Food Technology option
  Syllabus 9525: Module 1060: Biochemistry
              Module 1063: Food Technology
University of London Examinations and Assessment Council:
  Nuffield: Biochemistry
  Nuffield: Food science
University of Oxford Delegacy of Local Examinations:
  Module 4 Option: Biochemistry
Oxford and Cambridge Schools Examination Board:
  Biochemistry Module

Before embarking on an optional topic such as *Biochemistry and Food Science*, students will have completed the core modules, covering atomic structure, the chemical bond and a firm foundation of physical, inorganic and organic chemistry. Should they need to revise this core material, they can consult the references to *ALC*, which are to my text, *A-level Chemistry*, Third Edition (Stanley Thornes). They give the section of this text in which the relevant core material can be found. Students who are using a different A-level textbook need to consult the index of their book to find the corresponding material.

Acknowledgements

I thank Dr J.M. Gregory for reading the first draft and making many valuable comments.

I thank the following for supplying photographs:

Martyn Chillmaid: Figures 14.3A–D
Range Pictures Ltd: Figure 5.2F (The Bettmann Archive)
Science Photo Library: Front cover (Oxford Molecular Biophysics Laboratory);
Figure 5.2A (Science Source); Figure 10.3F (CNRI)
Tweedy of Burnley: Figure 11.3B

My family have given me their support and encouragement all through the writing of this book.

E.N. Ramsden
Oxford, 1995

Part 1

BIOCHEMISTRY

Biochemistry is the study of the chemical compounds which constitute living organisms, of the functions which these compounds perform and of the chemical reactions which take place during the life processes of the organism.

The questions which occupy the minds of biochemists can be summarised in four categories:

● What are the chemical compounds that occur in living cells? How can they be separated and identified?
● How are these compounds synthesised? What reactions do they take part in within the cell? What relationships exist between the different compounds in the cell? What interconversions take place between them?
● How are the cell reactions regulated so that the cell maintains its structure and its activities?
● What makes cells of different tissues, organs and species different from one another?

Biochemistry is a rapidly expanding subject. In the 1930s and 1940s, the application of chemical kinetics elucidated metabolic pathways in which reactants are converted into products through a series of reactions which take place in sequence under the control of enzymes. In the 1950s the electron microscope became more widely available in laboratories and led to a rush of discoveries. Also in the 1950s, the application of X-ray crystallography led to the discovery of the structure of DNA, the genetic material. In the 1970s and 1980s, a range of new techniques including biotechnology and genetic engineering have opened up exciting new fields of research and important new medical and industrial applications.

1

CELLS

1.1 STRUCTURE

All living things are made of **cells**. Some organisms, e.g. bacteria, consist of single cells. Plants and animals are multicellular organisms. The cells of which they are composed are different but share many similar features. Early studies of the structure of cells were done with the light microscope, which has a magnification of up to 1500 and a resolving power of about 200 nm (that is, it will distinguish objects that are 200 nm apart). Thanks to the invention of the electron microscope, which gives a magnification of about 500 000 and a resolution of about 1 nm, it is possible to see the structure of cells in great detail.

All living things are made of cells. The light microscope and the electron microscope have revealed the structure of cells in detail.

While a multicellular organism contains different sorts of cells, e.g. muscle cells, red blood cells, bone cells, nerve cells, skin cells, all these cells have many features in common. They all contain a fluid called **cytoplasm**, which consists of an aqueous medium, **cytosol**, and various structures. The cells are all bounded by a semi-permeable membrane. All cells contain structures; those that are surrounded by membranes are called **membranous organelles**. Some cells secrete surrounding walls, e.g. plant cells have a wall of cellulose.

Living cells contain ions, e.g. Na^+, K^+, Ca^{2+}, HCO_3^- and molecules of simple organic compounds such as butanedioic acid (succinic acid) and propane-1,2,3-triol (glycerol). They also contain large molecules with relative molecular masses from tens of thousands to several million. They will be described in the following sections.

Figures 1.1A and 1.1B are generalised pictures of a plant cell and an animal cell. Features of different types of plant and animal cell viewed under an electron microscope have been combined to give these generalised pictures. The functions of the structures shown in the cells are listed in Table 1.1A.

FIGURE 1.1A
A Generalised Plant Cell

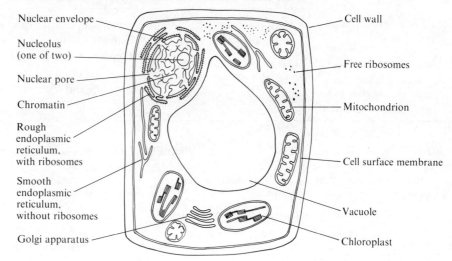

Nuclear envelope

Nucleolus (one of two)

Nuclear pore

Chromatin

Rough endoplasmic reticulum, with ribosomes

Smooth endoplasmic reticulum, without ribosomes

Golgi apparatus

Cell wall

Free ribosomes

Mitochondrion

Cell surface membrane

Vacuole

Chloroplast

The structures observed in a typical plant cell and a typical animal cell are illustrated.

Figure 1.1B
A Generalised Animal
Cell

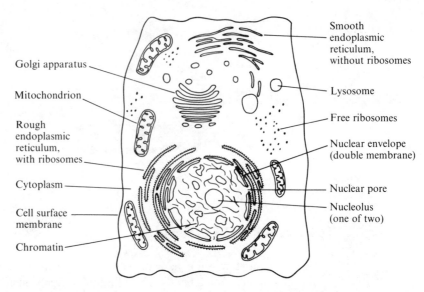

*The functions of a number
of cell structures are
tabulated.*

Structure	Function
Nucleus	Passes inherited characteristics from one generation to the next. Controls cell activities.
Cell surface membrane	Selectively permeable; allows small molecules and ions to pass into and out of the cell. Can also break to allow larger molecules to pass through and then reform.
Mitochondria	Sites of aerobic respiration, where compounds derived from carbohydrates, lipids and amino acids are oxidised, with the release of energy
Chloroplasts	Site of photosynthesis
Ribosomes	Site of protein synthesis
Lysosomes	Control of the destruction of cellular material and foreign material
Endoplasmic reticulum	A system of membranes which connect to form channels. On the surface of the endoplasmic reticulum are ribosomes. Proteins synthesised in the ribosomes cross the membrane of the endoplasmic reticulum and enter the channels, which form a transport system.

TABLE 1.1A
Cell Structures

CHECKPOINT 1.1

1. Describe the function of the following organelles within the cell: mitochondria, ribosomes, lysosomes

2. List four features which are common to plant and animal cells.

3. State two important features of plant cells which are absent in animal cells.

1.2 ENERGY

Energy is the capacity for doing work. It may derive from the position of an object – potential energy – or from the motion of an object – kinetic energy.

Energy is defined as the **capacity for doing work**. Heat, light, sound, electrical energy and the energy of chemical bonds are all forms of energy. The energy which an object possesses because of its position is called **potential energy**. The energy which an object possesses because it is moving is called **kinetic energy**. One form of energy can be converted into another.

The energy stored in chemical bonds is potential energy: it derives from the positions of atoms and ions relative to one another. The reactants and products in a chemical reaction contain different bonds and therefore different amounts of energy. In consequence, when the reactants change into the products, energy is either taken in or given out. Reactions in which energy is given out are **exothermic reactions** (also called exogenic reactions). Reactions in which energy is taken in are **endothermic reactions** (also called endogenic reactions). When a reaction takes place at constant pressure, as is the case with most biochemical reactions, the energy change which accompanies the reaction is called the change in **enthalpy** [see *ALC*, § 10.2]. Enthalpy has the symbol H, and change in enthalpy has the symbol ΔH. The **standard enthalpy change of reaction**, ΔH_R, is the heat **absorbed** in a reaction at one atmosphere pressure (1 atm) between the number of moles of reactants shown in the equation for the reaction. Notice that ΔH is the heat absorbed. If the reaction is exothermic, the value of ΔH is negative; if the reaction is endothermic, the value of ΔH is positive. In discussion of the energy values of foods, you will frequently meet the **standard enthalpy of combustion**, ΔH_C; this is the heat absorbed when 1 mole of a substance is completely burned in oxygen at 1 atm. Combustion is exothermic, so values of ΔH_C are negative.

An energy change that occurs at constant pressure is called an enthalpy change. Standard enthalpy change of reaction is defined.

Living organisms obey the **first law of thermodynamics** [see *ALC*, § 10.1.1]. This law states: In an isolated system the total energy remains constant. (A system is a part of the universe which has been selected for study in isolation from the rest of the universe. It may be a living organism and its immediate surroundings.) This law is another way of stating the **law of conservation of energy**: energy cannot be created or destroyed. A living organism may give energy to or take energy from its surroundings. A living organism may do work on its surroundings (e.g. expel a gas) or the surroundings may do work on the organism (e.g. compress it). All these exchanges must occur in such a way that the total energy of the organism and its surroundings remains constant.

Living organisms obey the laws of thermodynamics. Energy is neither created nor destroyed. Natural processes occur in the direction that increases entropy (disorder).

Another law of thermodynamics states that all natural processes proceed in a direction which increases the **entropy** of a system. Entropy is the degree of randomness, the degree of chaos in a system [see *ALC*, § 10.9]. A living organism is a system of low entropy. Living systems take in substances from their environment and build them into structures, thus decreasing their entropy. They release substances to the environment, thereby increasing the entropy of the environment. The total entropy of (living system + environment) increases.

CHECKPOINT 1.2

1. Why can the energy stored in chemical bonds be described as 'potential energy'?

2. The standard enthalpy of combustion of glucose is $-2830 \, \text{kJ mol}^{-1}$.

(*a*) What is meant by the terms (i) enthalpy, (ii) standard enthalpy, (iii) standard enthalpy of combustion?

(*b*) What does the negative sign in $-2830 \, \text{kJ mol}^{-1}$ mean?

3. When one gram of water vaporises, does its entropy increase or decrease?

4. When 20 amino acids combine to form a peptide, does the entropy of this matter increase or decrease?

1.3 FOOD

Later sections of this book will deal with the substances which human beings need to live. These are the carbohydrates, fats and proteins which provide energy, the proteins which build new tissues and repair old tissues and minerals and vitamins. Table 1.3A lists the recommended daily intake of these nutrients for people of different kinds.

The recommended daily intake of some nutrients is listed for reference later in the text.

Sex, age	Energy /mJ	Protein /g	Iron /mg	Calcium /mg	Vitamin C /mg	Thiamin /mg	Nicotinic acid/mg
0–1 year	6	35	7	600	20	0.6	7
Boys, 15–17	12	72	12	600	30	1.2	19
Girls, 15–17	9	53	12	600	30	0.9	19
Men, 18–55: moderately active very active	12 14	72 84	10 10	500 500	30 30	1.2 1.3	18 18
Women, 18–55: most occupations pregnancy	9 10	54 60	12 13	500 1200	30 60	0.9 1.0	15 18

TABLE 1.3A
Recommended Daily
Intake of some Nutrients

2

PROTEINS

2.1 FUNCTION

Proteins are essential for the life of all living organisms. Some of the functions which they perform are as follows:

- **Structural proteins** form the main part of cartilage, skin, hair, horns, hooves and feathers.
- **Contractile proteins**, e.g. myosin and actin of muscles, enable animals to move.
- **Enzymes** and many of the **hormones** which control metabolism in animals are proteins.

The functions of different types of proteins are listed.

- **Transport proteins**, e.g. haemoglobin, carry vital substances through the organism.
- **Immunoproteins** are antibodies which bind to substances which are foreign to a mammal.

The important functions of proteins gave them their name, which comes from the Greek word for *primary*.

Proteins include **simple proteins** and **conjugated proteins**. The latter consist of proteins combined with another substance. In shape protein molecules may be fibrous or globular. [See Table 2.1A.]

Simple proteins	
Fibrous proteins:	Insoluble in water and dilute solutions of salts. They include:
Collagens:	Present in cartilage, bone and tendons Strong, inelastic Converted by prolonged boiling water into gelatin, which is soluble in hot water
Elastins:	Present in artery walls, skin and tendons Elastic Unchanged by boiling in water
Globular proteins:	Molecules are approximately spherical.
Albumins:	Soluble in water; easily coagulated by heat Present in milk, egg white, blood plasma and muscle cells
Conjugated (complex) proteins:	Contain proteins combined with other compounds.
They include: Phosphoproteins: Glycoproteins: Lipoproteins: Chromoproteins: Nucleoproteins:	 Protein + phosphate Protein + carbohydrate Protein + lipid (fat or oil) Protein + a coloured group Protein + a nucleic acid

The characteristics of simple proteins (fibrous and globular) and complex proteins are listed.

TABLE 2.1A
Characteristics of
Proteins

All proteins are compounds of carbon, hydrogen, oxygen and nitrogen and some also contain sulphur and phosphorus.

2.2 FORMULAE

Proteins of these different kinds are all composed of the same structural units or 'building blocks': the amino acids, of which there are about twenty. Amino acids have the structure:

Amino acids are the structural units from which proteins are assembled.

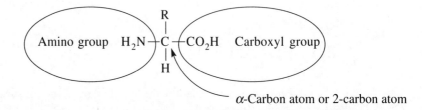

α-Carbon atom or 2-carbon atom

The general formula for an α-amino acid is $H_2NCHRCO_2H$.

The amino acids found in proteins have the amino group attached to the α-carbon atom (or 2-carbon atom); that is, the carbon atom next to the carboxyl group, and are therefore named α-amino acids. Each α-amino acid has a different R group [see Table 2.2A].

The group R may be

- non-polar and hydrophobic (water-hating), e.g. H in glycine, CH_3 in alanine, $CH(CH_3)_2$ in valine
- uncharged and hydrophilic (water-loving), e.g. —CH_2OH in serine, —CH_2SH in cysteine
- negatively charged at pH 6–7, e.g. —CH_2CO_2H in aspartic acid, —$(CH_2)_2CO_2H$ in glutamic acid
- positively charged at pH 6–7, e.g. —$(CH_2)_4NH_2$ in lysine

The α-carbon atom is asymmetric (see *ALC*, § 5.1.4) in all α-amino acids except glycine. There are two possible stereoisomers. In the case of alanine, $H_2NCH(CH_3)CO_2H$, these can be shown as in Figure 2.2A.

FIGURE 2.2A
L-Alanine and D-Alanine

Amino acids (except glycine) exist as D- and L-stereoisomers.

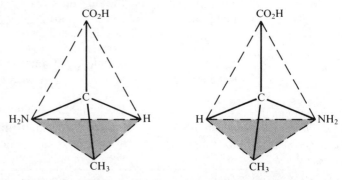

The stereoisomer with the amino group on the left when viewed in this way, with the carboxyl group at the top of the tetrahedron, is called L-alanine, while its mirror image is called D-alanine. You will visualise the difference better if you construct models. The same nomenclature applies to amino acids with a side-chain other than the CH_3 group of alanine. In each case, one of the isomers rotates the plane of plane-polarised light to the left and the other to the right, but the letters D- and L- do not indicate which direction this will be. All the naturally occurring amino acids found in proteins are of the L-configuration.

Amino acids are crystalline solids with melting temperatures above 200 °C. They dissolve more readily in water than in organic solvents. Why is there this big difference in properties between ethanoic acid CH_3CO_2H, and α-aminoethanoic acid, $H_2NCH_2CO_2H$? The explanation is the dipolar ions, **zwitterions**, formed by amino acids. Both the $—NH_2$ group and the $—CO_2H$ group can ionise. In aqueous solution, an amino acid can act

either (1) as an acid, a proton donor:

$$H_2N-\underset{\underset{H}{|}}{\overset{\overset{R}{|}}{C}}-CO_2H(aq) \rightleftharpoons H_2N-\underset{\underset{H}{|}}{\overset{\overset{R}{|}}{C}}-CO_2^-(aq) + H^+(aq)$$

or (2) as a base, a proton acceptor:

Amino acids are ionic, crystalline solids. Both the $—NH_2$ group and the $—CO_2H$ group can ionise. The dipolar ion, zwitterion, has no overall charge at a certain pH, called the isoelectric point.

$$H_2N-\underset{\underset{H}{|}}{\overset{\overset{R}{|}}{C}}-CO_2H(aq) + H_2O(l) \rightleftharpoons H_3N^+-\underset{\underset{H}{|}}{\overset{\overset{R}{|}}{C}}-CO_2H(aq) + OH^-(aq)$$

The higher the pH, the more ionisation (1) occurs. The lower the pH, the more ionisation (2) occurs. There is a certain pH at which the two ionisations are exactly balanced, and the amino acid is entirely in the form of the zwitterion,

$$H_3N^+-\underset{\underset{H}{|}}{\overset{\overset{R}{|}}{C}}-CO_2^-$$

This pH is called the **isoelectric point**. There is no net electric charge on the amino acid at this pH, and the dipolar ion will not move in an electric field.

Table 2.2A lists α-amino acids of formula $H_2NCHRCO_2H$.

In addition, proline and hydroxyproline occur in proteins:

$$\begin{array}{c} HN-CH-CO_2H \\ {/} \qquad {\backslash} \\ CH_2 \qquad CH_2 \\ {\backslash} \quad {/} \\ CH_2 \end{array}$$
Proline

$$\begin{array}{c} HN-CH-CO_2H \\ {/} \qquad {\backslash} \\ CH_2 \qquad CH_2 \\ {\backslash} \quad {/} \\ CH \\ | \\ OH \end{array}$$
4-Hydroxyproline

An amino acid contains an acidic $—CO_2H$ group and a basic $—NH_2$ group. When the carboxyl group of one amino acid reacts with the amino group of a second amino acid, a molecule of water is eliminated and a **peptide** is formed.

Many amino acids may combine to form polypeptides and proteins.

The peptide which is formed contains two amino acid residues joined by the group $—CONH—$, which is called the **peptide link**. Since it contains a carboxyl group and an amino group, the peptide can combine with other amino acids or peptides through the formation of more peptide bonds and the elimination of water. Many amino acids may combine to form a peptide (2–20 amino acid residues), a **polypeptide** (20–50 amino acid residues) or a **protein** (50–1000 amino acid residues). Peptides and proteins can be described as **condensation polymers** of amino acids.

General structure:
$$H_2N-\overset{\overset{\displaystyle H}{|}}{\underset{\underset{\displaystyle R}{|}}{C}}-CO_2H$$

Name	Abbreviated name	R	Essential amino acids (E)
Glycine	Gly	—H	
Alanine	Ala	—CH_3	
Valine	Val	—$CH(CH_3)_2$	E
Leucine	Leu	—$CH_2CH(CH_3)_2$	E
Isoleucine	Ileu	—$CH(CH_3)(CH_2CH_3)$	E
Serine	Ser	—CH_2OH	
Threonine	Thr	—$CH(OH)CH_3$	E
Aspartic acid	Asp	—CH_2CO_2H	
Asparagine	Asn	—CH_2CONH_2	
Glutamic acid	Glu	—$CH_2CH_2CO_2H$	
Glutamine	Gln	—$CH_2CH_2CONH_2$	
Lysine	Lys	—$(CH_2)_4NH_2$	E
5-Hydroxylysine		—$(CH_2)_2CH(OH)CH_2NH_2$	
Arginine	Arg	—$(CH_2)_3NHC(NH_2)NH_2$	
Cysteine	CysH	—CH_2SH	
Cystine	Cys—Cys	—$CH_2SSCH_2CH(NH_2)CO_2H$	
Methionine	Met	—$(CH_2)_2SCH_3$	E
Phenylalanine	Phe	—CH_2⟨phenyl ring⟩	E
Tyrosine	Tyr	—CH_2⟨phenyl ring⟩—OH	
Tryptophan	Try	—CH_2—C (indole ring) HC, NH	E
Histidine	His	—CH_2—C=CH, HN, N, C, H (imidazole ring)	E (for infants)

Naturally occurring amino acids are listed.

TABLE 2.2A
Amino Acids Found in Proteins

$$H_2N-\underset{\underset{H}{|}}{\overset{\overset{R^1}{|}}{C}}-\overset{\overset{O}{\diagup\!\!\diagup}}{\underset{\diagdown}{C}}_{O-H} \quad + \quad H_2N-\underset{\underset{H}{|}}{\overset{\overset{R^2}{|}}{C}}-\overset{\overset{O}{\diagup\!\!\diagup}}{\underset{\diagdown}{C}}_{O-H}$$

The —CO$_2$H group of one amino acid can react with the —NH$_2$ group of another amino acid to form a peptide and water. The peptide link is —CONH—.

$$H_2N-\underset{\underset{H}{|}}{\overset{\overset{R^1}{|}}{C}}-\overset{\overset{O}{||}}{C}-\underset{\underset{H}{|}}{N}-\underset{\underset{H}{|}}{\overset{\overset{R^2}{|}}{C}}-\overset{\overset{O}{\diagup\!\!\diagup}}{\underset{\diagdown}{C}}_{O-H} \quad + H_2O$$

The peptide link

2.3 PROTEIN IN THE DIET

Proteins can be hydrolysed to amino acids by strong mineral acids, e.g. 6 mol dm^{-3} hydrochloric acid. In the body, proteolytic enzymes convert proteins into amino acids. These amino acids can then be used by the body to synthesise the proteins which it needs. Many amino acids can be synthesised in the body. There are, however, eight amino acids which cannot be synthesised and which must therefore be ingested (see those marked E in Table 2.2A). Young children also need histidine.

A diet which contains the eight essential amino acids provides the adult body with the materials necessary to synthesise all the amino acids and proteins which it needs.

Animals synthesise proteins from amino acids. Some amino acids are synthesised in the body, while others are obtained from proteins in the diet. Eight essential amino acids cannot be synthesised in the body and must be part of the diet.

The quality of food protein is judged not only on its protein content but also on the number of essential amino acids which it contains. The ease of digestion and absorption of the protein food is also important. The protein quality of foods is rated relative to egg protein. Eggs score 100%, milk 95%, beef 93%, soya beans 74%, rice 67% and maize 49%. A diet should contain a variety of proteins so that essential amino acids which are lacking in one protein are supplied by another. For a mixed diet, the daily requirement of protein is 53 g for the average man and 41 g for the average woman [see Table 1.3A].

2.4 HOW TO TEST FOR PROTEIN

Compounds which contain peptide bonds give a characteristic purple colour when treated with an alkaline solution of copper(II) sulphate. The test is called the **biuret test** because it is given by the compound biuret, $H_2NCONHCONH_2$, which contains a peptide link. The colour ranges from pink to violet as the number of peptide bonds in the test substance increases, and proteins give a deep blue-violet colour.

The biuret test is used to detect protein.

2.5 MEASURING PROTEIN CONTENT

The protein content of a food can be found by the Kjehldahl method of converting the nitrogen into ammonia and estimating the ammonia.

The proteins in a food item are the only significant source of nitrogen. If the nitrogen content of a food can be found, this value will give an indication of the protein content. The Kjehldahl method is used to find the nitrogen content. The substance is boiled with concentrated sulphuric acid to convert the nitrogen into ammonium sulphate. This is done in a fume cupboard in a Kjehldahl flask, a round-bottomed flask with a long neck. When the reaction is complete, the mixture is allowed to cool and transferred to a distillation flask. An excess of alkali is added to liberate ammonia from ammonium sulphate, and the ammonia is distilled into a known volume of standard acid. The excess of acid is found by titration against a standard alkali. By subtraction, the volume of acid used to neutralise ammonia is found. The percentage of nitrogen in the food sample can then be calculated.

Worked example   A mass of flour $= 2.0\,g$ was treated by the Kjehldahl method, and ammonia was distilled into $50\,cm^3$ of $0.10\,mol\,dm^{-3}$ sulphuric acid. The excess of acid needed $40\,cm^3$ of $0.10\,mol\,dm^{-3}$ sodium carbonate solution for neutralisation, using methyl orange as indicator.

Since $40\,cm^3$ of $0.10\,mol\,dm^{-3}$ Na_2CO_3 neutralise $40\,cm^3$ of $0.10\,mol\,dm^{-3}$ H_2SO_4, $10\,cm^3$ of $0.10\,mol\,dm^{-3}$ H_2SO_4 were used by the NH_3.

Amount of $NH_3 = 10 \times 10^{-3} \times 0.10 \times 2\,mol = 0.0020\,mol$

Therefore amount of $N = 0.0020\,mol$ and mass of $N = 0.028\,g$

Percentage of N in flour sample $= 0.028 \times 100/2.0 = 1.4\%$

The percentage of protein can be found approximately by multiplying the percentage of nitrogen by the factor 6.25. The factor 6.25 is an average value: the percentage of nitrogen in protein varies between different proteins.

Percentage of protein in flour sample $= 6.25 \times 1.4\% = 8.8\%$

CHECKPOINT 2.5

1. Refer to the formulae of amino acids in Table 2.2A.

(*a*) What is the side-chain in aspartic acid?

(*b*) How will an increase in pH affect the degree of ionisation of the side-chain of aspartic acid?

(*c*) What is the side-chain in lysine?

(*d*) How will a decrease in pH affect the degree of ionisation of the side-chain of lysine?

(*e*) Why might a change in pH affect the overall shape of a protein?

2. Aspartame is an artificial sweetener which is 160 times sweeter than sucrose. It is a dipeptide of the amino acids aspartic acid and phenylalanine, with the methyl ester of phenylalanine forming the C-terminal end.

Draw the structure of aspartame. (Refer to Table 2.2A for the formulae of the amino acids).

2.6 PRIMARY STRUCTURE OF PROTEINS

2.6.1 PURIFYING PROTEINS

Proteins can be precipitated from solution by adjusting the pH and the concentration of ions.

Before taking steps to find the structure of a protein, the protein must first be obtained pure. It is precipitated from solution by adding a salt, e.g. ammonium sulphate. The precipitate is collected by centrifugation. Like amino acids, a protein has an isoelectric point. Most proteins have minimum solubility near the isoelectric point. The pH is adjusted to achieve precipitation of the protein of interest and leave others in solution.

A protein may be hydrolysed to amino acids by boiling with a concentrated solution of hydrochloric acid.

In order to determine the structure of a polypeptide or a protein, it is necessary to know which amino acids are present. Hydrolysis of the peptide bonds is carried out by heating the polypeptide with $6 \, mol \, dm^{-3}$ hydrochloric acid at $100 \, °C$ to $120 \, °C$ for 10–24 hours. The amino acids are then separated by one of the following methods.

2.6.2 METHODS OF SEPARATING AMINO ACIDS

PAPER CHROMATOGRAPHY

The mixture of amino acids can be separated...

...by paper chromatography, using ninhydrin to detect the positions of the amino acids in the chromatogram...

The mixture of amino acids is applied to a sheet of chromatography paper [see *ALC*, §8.7.3]. The paper is supported on a glass plate and stood in a glass tank containing the developing solvent, which is usually butan-1-ol, ethanoic acid and water. As the solvent ascends the sheet of chromatography paper, different amino acids move at different rates. When the solvent reaches nearly to the end of the paper, the paper is lifted out of the tank. The position of the solvent front is marked, and the paper is dried. The amino acids are colourless, and the chromatography paper looks blank. Ninhydrin detecting reagent (a solution of ninhydrin in propanone) must be used to reveal the amino acids. When the chromatogram is sprayed with ninhydrin solution and heated, the different amino acids appear as coloured spots. They can be identified from their R_F values (the distance they have moved relative to the solvent front).

...or by thin layer chromatography

Descending paper chromatography can also be used, and thin layer chromatography can be used [see *ALC*, §8.7.4].

ION EXCHANGE CHROMATOGRAPHY

...or by ion exchange chromatography, using a negatively charged resin to adsorb cationic amino acids and a positively charged resin to adsorb negatively charged amino acids.

A chromatographic column is filled with granules of a synthetic resin which contains charged groups, either positively charged groups (in anion-exchange resins) or negatively charged groups (in cation-exchange resins). For a mixture of amino acids, a cation-exchange column filled with sulphonated poly(styrene) resin may be used. If the separation is carried out at pH 3, the sulphonic acid groups in the resin are ionised, and most of the amino acids are cations. The most basic amino acids, e.g. lysine, histidine, are bound most tightly, and the most acidic amino acids, e.g. glutamic acid, aspartic acid, are bound least tightly. When an **eluant**, e.g. aqueous sodium chloride, is passed down the column, some amino acids are displaced more readily than others and are the first to flow out of the column in the **eluate**. The procedure has been automated in an apparatus called an **amino acid analyser**. It enables the identities and the amounts of all the amino acids present in a protein to be found in a few hours, using less than a milligram of protein.

GEL PERMEATION CHROMATOGRAPHY

Gel permeation chromatography separates proteins and peptides according to size.

Another type of column chromatography is **gel permeation chromatography**. It uses as stationary phase a gel which contains particles consisting of a network of polysaccharide fibres. A solution of proteins is applied to the column. The largest protein molecules cannot penetrate the pores in the particles and pass quickly down the column. Smaller protein molecules can enter some of the larger pores. The smallest protein molecules can penetrate the smallest pores in the particles. The mixture of solutes is washed from the column by an eluant. Separation occurs according to molecular size [see Figure 2.6A].

Calcium phosphate gel, alumina gel, starch and hydroxyapatite are used as adsorbents. Different gels are available which allow the separation of proteins with molar masses ranging from a few hundred to over a hundred million. The greatest separation is achieved by using very small gel particles, but the rate of flow through the column is then very slow.

FIGURE 2.6A
Gel Permeation
Chromatography

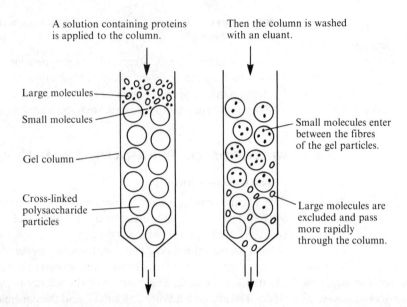

A solution containing proteins is applied to the column.

Then the column is washed with an eluant.

Large molecules

Small molecules

Gel column

Cross-linked polysaccharide particles

Small molecules enter between the fibres of the gel particles.

Large molecules are excluded and pass more rapidly through the column.

ELECTROPHORESIS

Electrophoresis separates amino acids on the basis of their overall charge and their charge/mass ratio.

The mixture of amino acids is put on to chromatography paper or a gel and a potential difference is applied. The amino acids migrate towards the positive or negative electrode at a speed which depends on the ratio of charge/mass. The location of the amino acids is shown by, for example, spraying with ninhydrin (see paper chromatography).

2.6.3   THE SEQUENCE OF AMINO ACIDS

Amino acid analysis is the first step in elucidating the structure of a protein. Proteins have relative molecular masses of 5000–36 000. A molecule may contain more than one polypeptide chain, and an individual polypeptide chain contains 100–300 amino acid residues. After finding out which amino acid residues are present, the biochemist sets about finding out the order in which several hundred amino acid residues are linked by peptide bonds to form the polypeptide chain.

The sequence of amino acids in the protein must be mapped out. The first step is to find out which amino acid has the terminal —NH₂ group.

There is a useful reagent which tells the biochemist the identity of the **N-terminus** (the amino acid residue with a free —NH_2 group at the end of the chain). 1-Fluoro-2,4-dinitrobenzene is yellow, and it reacts with an amino group to form a yellow compound.

$$O_2N-\bigcirc-F \;+\; H_2N-Protein \;\longrightarrow\; O_2N-\bigcirc-\overset{\overset{\displaystyle H}{|}}{N}-Protein \;+\; HF$$

NO₂

NO₂

1-Fluoro-2,4-dinitrobenzene

Yellow compound

Then chromatography will identify the amino acid which moves as a yellow spot.

The sequence of amino acids in the polypeptide chain is worked out by partially hydrolysing the polypeptide to give small peptide fragments. The sequence of amino acids in the peptide fragments is worked out by further hydrolysis and analysis. By looking for overlaps between sequences of amino acid groups in the peptide fragments, the biochemist gradually assembles a picture of the whole chain. For example, the results of an analysis might be:

N-terminus: His

Fragments obtained by hydrolysis 1:　Asp — Tyr — Glu — Leu — Arg
　　　　　　　　　　　　　　　　　His — Lys
　　　　　　　　　　　　　　　　　Gly — Ala

Fragments obtained by hydrolysis 2:　His — Lys — Asp — Tyr
　　　　　　　　　　　　　　　　　Glu — Leu — Arg — Gly — Ala

Partial hydrolysis of the protein yields peptides. By overlapping the fragments, the amino acid sequence can be deduced. This is the primary structure of the protein.

From these results can be deduced the sequence:

His — Lys — Asp — Tyr — Glu — Leu — Arg — Gly — Ala.

You can imagine the meticulous labour involved in completing the sequence for a protein consisting of several hundred amino acids. This sequence is called the **primary structure** of the protein.

2.7 SECONDARY STRUCTURE OF PROTEINS

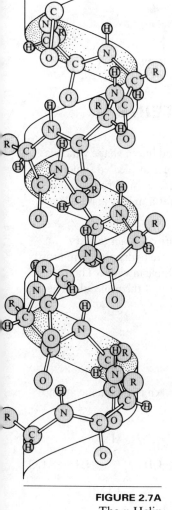

X-ray diffraction studies show that in many protein molecules there is a regular arrangement of sections of the polypeptide chains. The α-helix is one such arrangement. It is a regular spiral with 3.6 amino acids per turn (18 amino acids in 5 spirals). The R groups point towards the outside of the helix. The nitrogen atom in each peptide group forms a hydrogen bond to the oxygen atom of a peptide group further along the helix. The hydrogen bonds ⊃N—H ⋯ O=C⊂ are linear and therefore of maximum strength. They hold the peptide chain in its helical form [see Figure 2.7A]. An α-helix can form with either L- or D-amino acids but not with a mixture of both. From the naturally occurring L-amino acids can be formed a right-handed or a left-handed coil. All the α-helices found in proteins are right-handed (turn clockwise as you move along the chain away from you). [See also *ALC*, §4.7.3, Figure 4.39.]

The *a*-helix is the **secondary structure** of the protein. It is a flexible and elastic structure which occurs in, for example, the fibrous α-keratin proteins in hair and wool.

Many protein molecules adopt an α-helical conformation. This is the secondary structure of the protein.

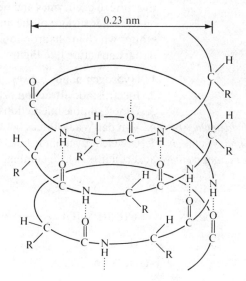

FIGURE 2.7A
The α-Helix

The β-pleated sheet is another type of secondary structure of proteins. Polypeptide chains are aligned side by side and are connected by hydrogen bonds between all the peptide links [see Figure 2.7B]. The β-pleated sheet is flexible but inelastic. It is the structure of the protein β-keratin present in silk fibres.

FIGURE 2.7B
The β-Pleated Sheet

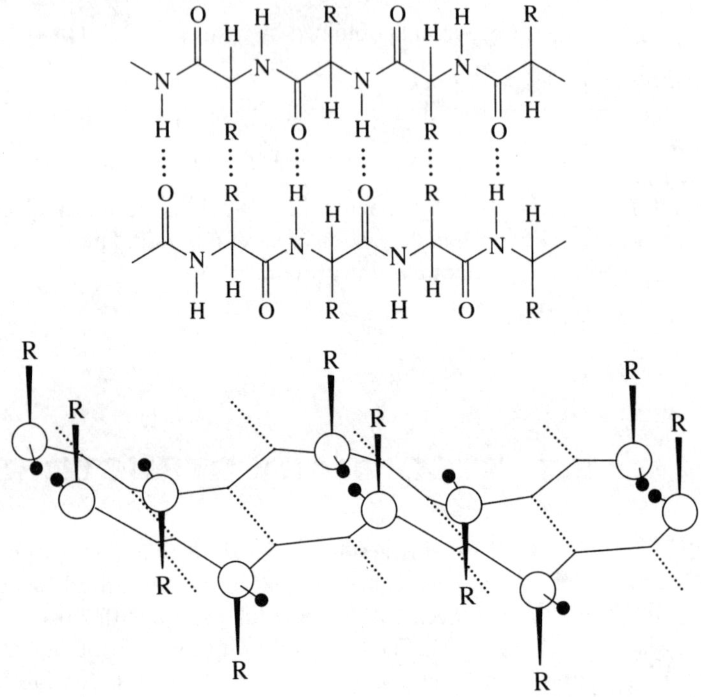

Other proteins adopt a β-pleated sheet as their secondary structure.

2.8 TERTIARY STRUCTURE OF PROTEINS

α-Helical structures and β-pleated structures fold and coil into a shape called the tertiary structure of the protein...

The chains or sheets of a protein are further folded and coiled into a shape which is called the **tertiary structure** of the protein. The helical structure or pleated structure is partially interrupted in the folding to form the tertiary structure. In fibrous proteins, long molecules coil round each other to form fibres, e.g. keratin in hair and collagen in tendons. These proteins are insoluble.

In globular proteins, folding is more extensive, to give a more compact tertiary structure, e.g. enzymes and hormones. The R groups which are hydrophilic are mostly on the outer surface of the molecule, making the protein soluble in water. The R groups which are hydrophobic are mostly on an inner surface of the molecule. The tertiary structure [see Figure 2.8A] results from:

1. Hydrogen bonds between amino acids
2. Electrostatic attraction between polar groups
3. Dipole–dipole interactions
4. Van der Waals forces between non-polar groups

...which is held in place by chemical bonds between groups in the peptide chain.

5. Covalent disulphide links. These strong bonds form between two cysteine amino acid residues, which combine to form cystine.

$$2H_2N-\underset{\underset{CO_2H}{|}}{\overset{\overset{H}{|}}{C}}-CH_2SH + [O] \longrightarrow H_2N-\underset{\underset{CO_2H}{|}}{\overset{\overset{H}{|}}{C}}-CH_2-S-S-CH_2-\underset{\underset{CO_2H}{|}}{\overset{\overset{H}{|}}{C}}-NH_2 + H_2O$$

Cysteine   Cystine   Disulphide bridge

FIGURE 2.8A
Interactions between Side-
chains in a Protein

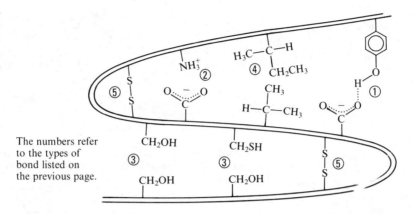

FIGURE 2.8A
Interactions between Side-
chains in a Protein

The numbers refer
to the types of
bond listed on
the previous page.

2.9 QUATERNARY STRUCTURE OF PROTEINS

Some protein molecules consist of a number of closely linked subunits. Some consist of protein and non-protein components. The association of subunits is called the quaternary structure.

Many protein molecules consist of two or more polypeptide chains. These are described as **oligomeric proteins**. Haemoglobin is an oligomeric protein. The molecule consists of four subunits. These are polypeptide chains of two kinds, which are described as α and β. The two α-polypeptide chains have a slightly different sequence of amino acid residues from the two β-polypeptide chains. The four subunits are closely associated to form a stable globular protein. The manner in which the subunits are grouped to form the protein molecule is the **quaternary structure** of the protein [see Figure 2.9A]. The term is also used to describe the binding of a polypeptide to a non-protein component in a protein structural unit.

FIGURE 2.9A
The Quaternary Structure
of Haemoglobin

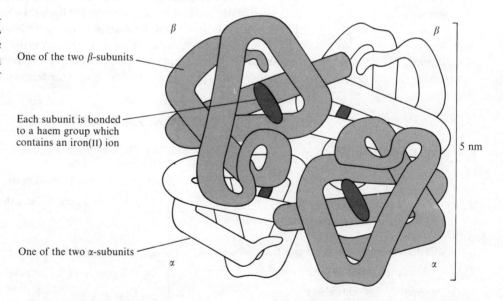

One of the two β-subunits

Each subunit is bonded
to a haem group which
contains an iron(II) ion

One of the two α-subunits

5 nm

The iron(II) ions in the haem groups bind oxygen and enable haemoglobin to perform its function of transporting oxygen round the body. There are forces of attraction between the α- and the β-subunits of the same kinds as those that maintain the tertiary structure (see above).

The structural protein collagen owes its strength to its superhelical quaternary structure.

Collagen is a fibrous structural protein. It constitutes about one-quarter of the total protein in mammals. It is the fibrous component of skin, bone, tendon, cartilage, blood vessels and teeth. Collagen is in the form of insoluble fibres of high tensile strength. The quaternary structure is a superhelix as shown in Figure 2.9B. Numbers of superhelices associate to form collagen fibres.

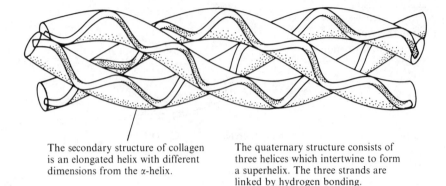

FIGURE 2.9B
The Quaternary Structure
of Collagen

The secondary structure of collagen
is an elongated helix with different
dimensions from the α-helix.

The quaternary structure consists of
three helices which intertwine to form
a superhelix. The three strands are
linked by hydrogen bonding.

2.10 DENATURATION OF PROTEINS

*The secondary, tertiary
and quaternary structures
of a protein can be
disrupted by unfavourable
conditions: the protein is
then denatured. High
temperature and high
pressure cause irreversible
denaturation. Urea causes
reversible denaturation.*

Changes in the secondary, tertiary and quaternary structure of proteins can occur if
conditions are allowed to vary outside a limited range of pH and temperature. When
most globular proteins are subjected to temperatures above 60 °C to 70 °C they become
insoluble and they lose their biological activity. Egg white coagulates on heating. The
primary structure of the protein remains but the secondary, tertiary and quaternary
structures are disrupted. The process is called **denaturation**. Changes in pH alter the
ionisation of the R side-chains and may break the electrostatic attractions which
maintain the tertiary and quaternary structure. At the pH where the ionisation of R
side-chains is such as to minimise the repulsion between protein molecules, the
molecules coagulate and precipitate. The change brought about by high temperature or
high pH or low pH is **irreversible denaturation**. The **reversible denaturation** of proteins is
brought about by a concentrated solution of urea (8 mol dm^{-3}). Hydrogen bonds form
between the polypeptide chain and urea, $CO(NH_2)_2$. The reduction in hydrogen
bonding between polypeptide chains destabilises the tertiary and quaternary structure.

CHECKPOINT 2.10

1. Describe the difference between

(a) a simple protein and a conjugated protein

(b) a globular protein and a fibrous protein

(c) a dipeptide and a polypeptide

(d) the primary structure and the tertiary structure of a
protein

(e) α-helix and β-pleated sheet configurations

2. Give a brief explanation of the following.

(a) Globular and fibrous proteins differ in solubility.

(b) Blood proteins help to protect against changes in pH.

(c) A protein has minimum solubility at its isoelectric point.

(d) Amino acids have high melting temperatures compared
with other molecules of the same size.

(e) The keratins of wool are flexible while those of hooves
are rigid.

(f) It is possible to give a 'permanent' wave to hair.

3. A peptide has the amino acid sequence shown below.

$$Ala - Cys - Asp - Ser$$
$$\qquad\qquad\quad |$$
$$N\ terminus\quad Val - Cys - Gly - Ala$$

(a) State the two kinds of linkages present.

(b) Draw the structural formula (Refer to Table 2.2A for the
formulae of the amino acids.)

2.11 ENZYMES

Enzymes are the substances that **catalyse** biological reactions. A **catalyst** is able to
speed up a chemical reaction without being used up in the reaction. It does this by

providing an alternative reaction route which has a lower activation energy [see *ALC*, § 14.10].

Enzymes can bring about reactions in aqueous solution, at the pH and temperature of living organisms. How can they bring about reactions under these mild conditions when many of the non-enzymic reactions we are familiar with require high temperature or high pressure or high or low pH to give a good yield? The decomposition of hydrogen peroxide is catalysed by the enzyme catalase. Under certain conditions, the enzyme has a **turnover number** of 5×10^6; that is, one molecule of the enzyme catalyses the reaction of 5×10^6 molecules of the substrate in one minute. This rate is higher by many powers of ten than that shown by other catalysts for the decomposition of hydrogen peroxide. This comparison illustrates the enormous catalytic power of enzymes.

Enzymes are the proteins that catalyse biological reactions. Compared with other catalysts, enzymes have enormous catalytic power and high turnover numbers.

Compared with the catalysts you have met in your study of chemical reactions [*ALC*, § 14.10], enzymes are very **specific**; that is, an enzyme catalyses the reactions of only one substance or a very limited range of substances. The substance which an enzyme enables to react is called the **substrate**. Most enzymes act within cells, and cells contain many substances. It is important that an enzyme targets its own substrate and leaves other substances untouched.

An enzyme is specific for a certain substrate or a group of similar substrates.

Often experiments with enzymes are done with extracts of biological material which are not pure. It is convenient to have a measure of the enzyme activity of the extract. The number of **enzyme units** present is the amount of substrate converted into product per unit time under specified conditions of pH etc. The unit is mole s^{-1}. **Specific enzyme activity** is the number of enzyme units per mg of enzyme protein. The unit is $\text{mol s}^{-1} (\text{mg protein})^{-1}$

The terms enzyme unit, specific enzyme activity and optimum temperature are defined.

The temperature dependence of enzyme-catalysed reactions is different from that of other reactions. The rate at first increases with temperature and then decreases as the enzyme is denatured. The temperature at which the enzyme shows maximum catalytic activity is called the **optimum temperature** for that enzyme.

2.12 ENZYME KINETICS

The kinetics of enzyme-catalysed reactions show that the rate of reaction is first order at low substrate concentration. As the substrate concentration increases, the reaction rate reaches a maximum and the order changes to zero.

Figure 2.12A shows the way in which the rate, V, of an enzyme-catalysed reaction increases with the concentration of the substrate. At low substrate concentration, the rate, V, is proportional to the substrate concentration: the reaction is first order with respect to the substrate [see *ALC*, § 14.5]. At high substrate concentration, the rate reaches a maximum, V_{max}: the reaction is zero order with respect to the substrate. At intermediate concentrations, the reaction is of mixed order. The value of V_{max} is proportional to the enzyme concentration.

FIGURE 2..12A
The Hydrolysis of ATP
Catalysed by Myosin

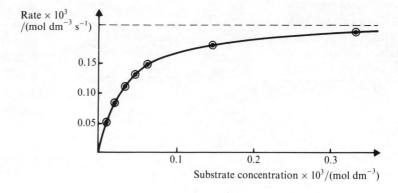

Michaelis and Menten suggested that an enzyme–substrate complex is formed by the adsorption of molecules of substrate on to active sites in the enzyme molecules.

The kinetics of enzyme-catalysed reactions led M. Michaelis and B. Menten to put forward a theory for the mechanism of enzyme-catalysed reactions. They suggested that enzyme and substrate combine reversibly to form an **enzyme–substrate complex**. This complex breaks down to form the enzyme and the product.

$$E + S \rightleftharpoons ES \quad \text{followed by} \quad ES \rightarrow E + P$$

They suggested that an enzyme possesses one or more **active sites** where the substrate is adsorbed. When the concentration of substrate is low, V increases linearly with [S] as substrate molecules fill the available active sites on the enzyme molecules. When the concentration of substrate is high enough to occupy all the available active sites, the enzyme is saturated with substrate, and the reaction rate reaches a maximum value. This rate, V_{max}, is called the **limiting velocity** or the maximum velocity at infinite substrate concentration. The substrate concentration at which $V = V_{max}/2$ is called the **Michaelis constant**, K_m.

The limiting velocity and the Michaelis constant are defined.

The value of the Michaelis constant, K_m, can be found by making measurements of the initial reaction velocity at different substrate concentrations with a fixed concentration of enzyme. The values of K_m and V_{max} for an enzyme vary with the structure of the substrate, with pH and with temperature.

2.13 THE ACTIVE SITE

2.13.1 LOCK AND KEY THEORY

Fischer originated the lock and key theory of the active site. He envisaged the active site as comprising a binding site and a catalytic site.

The **active site** of an enzyme is only a small region, perhaps 5% of the enzyme's surface. It is a crevice in the enzyme molecule into which the substrate molecule fits. The groups present at the active site have to bind the substrate and to catalyse the reaction. The fit between an enzyme and its substrate has been compared with that of a lock and a key. The **lock and key theory** of the active site was originated by Emil Fischer in 1894 [Figure 2.13A]. It takes the correct key to open a lock. It takes the correct enzyme to bond to the substrate and catalyse its reaction. The very precise fit is the reason why enzymes are so specific. The active site of an enzyme is composed of a **binding site** and a **catalytic site**. The binding between enzyme and substrate may involve electrostatic attraction, hydrogen bonding and van der Waals forces.

Lysozyme, with 129 amino acid residues, was the first enzyme for which the three-dimensional structure was worked out.

The first enzyme for which the three-dimensional structure was worked out was lysozyme. This antibacterial agent was discovered by Alexander Fleming in 1922. The reason for its antibacterial action is that it breaks open bacterial cells by hydrolysing polysaccharides in the cell walls. Figure 2.13B shows the sequence of 129 amino acid residues in lysozyme. X-ray crystallography revealed a three-dimensional structure containing helical regions and a pleated sheet region. The molecule is folded in such a way as to put residues with polar side-chains on the surface of the molecule and non-

FIGURE 2.13A
The Lock and Key Theory. The substrate binds to the active site of the enzyme and reacts

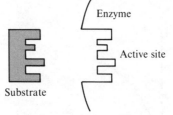

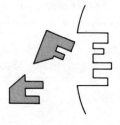

Substrate fits into active site. Reaction takes place to form the product or products.

The products leave the active site.

FIGURE 2.13B
Lysozyme – the sequence
of amino acid residues

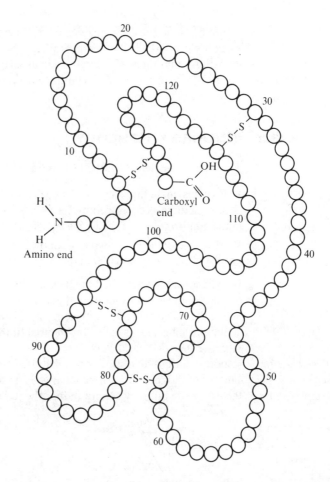

polar hydrophobic side-chains in the interior. The folding of the molecule results in the formation of a deep cleft which runs up one side of the molecule. This cleft is the active site of the enzyme. The substrate, the polysaccharide chain, fits into the cleft and binds to the enzyme by hydrogen bonding and other interactions [Figure 2.13C]. The active site is thought to include the carboxyl groups of glutamic acid (residue 35) and aspartic acid (residue 52).

FIGURE 2.13C
Lysozyme – the three-
dimensional structure

*The active site of lysozyme
has been identified as a
long deep cleft in the
molecule. The substrate, a
long polysaccharide
molecule, fits into the cleft
in the lysozyme molecule.*

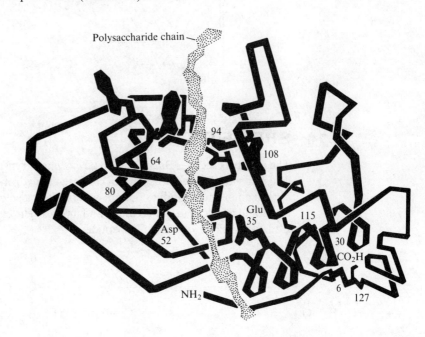

The catalytic site has polar groups which ease bond-breaking in the substrate.

The catalytic site of an enzyme interacts with the substrate in such a way as to weaken the bond which is to be broken. Often polar groups, such as the —NH_2 group in histidine and the —CO_2H group in aspartic acid, are the groups which catalyse the reaction. By attracting or repelling electrons in the substrate such groups can alter the strength of the bond which is to be broken.

2.13.2 INDUCED FIT THEORY

There are some defects in the lock and key theory of the active site.

● Proteins, including enzymes, have flexible conformations. Active sites need not be rigid structures as envisaged by Fischer. X-ray diffraction studies show that changes in the conformation of the active site occur when it binds the substrate.

● Molecules which are smaller than the substrate molecules and possess some reactive groups in common with it are not able to enter the active site and disrupt catalysis.

Koshland proposed the induced fit theory

according to which the substrate changes the conformation of the active site to achieve a perfect fit.

In 1958 Koshland modified the theory. He suggested that the shape of the active site of the enzyme is not exactly complementary to the shape of the substrate molecule. When the substrate binds to the enzyme, however, it changes the conformation of the active site. This idea is called the **induced fit theory**. Koshland's theory does not visualise specificity as a rigid fit of substrate and active site. It postulates a flexible response of the active site to the substrate [see Figure 2.13D]. Only the substrate is capable of inducing the correct change in an active site. Smaller molecules may enter the active site but they cannot induce the required change in its conformation.

FIGURE 2.13D
A Comparison of the Lock and Key Theory and the Induced Fit Theory

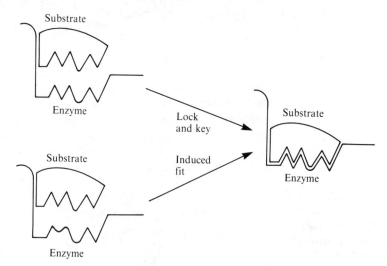

2.14 SPECIFICITY

The pancreas secretes three proteolytic enzymes: trypsin, chymotrypsin and elastase. The enzymes are **specific**: each can catalyse the hydrolysis of peptide links between certain amino acid residues but not others. They also catalyse the hydrolysis of certain esters and amides but not all esters and amides. The enzymes work best at neutral pH. Consider the hydrolysis of this peptide:

The nature of the groups R^1 and R^2 adjacent to the peptide link to be hydrolysed decides which enzyme will catalyse the reaction.

For hydrolysis by trypsin,

Proteolytic enzymes are specific for peptide bonds between certain amino acid residues.

- R^1 may belong to arginine or lysine. It must have an $-NH_3^+$ group.
- R^2 may be $-NH-$ or $-NH_2$ or $-OCH_3$

For hydrolysis by chymotrypsin,

- R^1 may be CH_3CO-, C_6H_5CO-, H_2NCH_2CO-, $C_6H_5CH_2OCO-$ or $-H$, that is, the NH_2 group may be free.
- R^2 may belong to tyrosine, phenylalanine, tryptophan or methionine

For hydrolysis by elastase,

- R^1 must belong to glycine or alanine.

Peptides can also be hydrolysed by an enzyme in gastric juice, pepsin, which has an optimum pH of 1.5–2.5. For pepsin,

- R^1 must belong to an aminodicarboxylic acid, e.g. glutamic acid
- R^2 must belong to an aromatic amino acid, and must not have a free $-NH_2$ group

Chymotrypsin has been the subject of much research. Evidence places the histidine residue at position 57 in the protein chain and serine (residue 195) and aspartate (residue 102) at its active site.

- The inhibitor DFP (see next paragraph) combines with serine (residue 195). The enzyme then loses its catalytic activity, showing that the serine group is part of the active site.
- The manner in which the rates of chymotrypsin-catalysed reactions depend on pH indicates that a group with the pK of a histidine residue is part of the active site.
- Other evidence places aspartic acid (residue 102) at the active site. At pH 7, the pH at which chymotrypsin works best, this is ionised as aspartate ion.

The conformation of the enzyme is thought to bring the histidine residue at position 57 into position to form a hydrogen bond with serine (residue 195). This enables the oxygen atom of the serine $-OH$ group to bind the substrate. After the substrate has bound to serine (residue 195), the histidine residue has a positive charge, and this is stabilised by interaction with aspartate ion.

FIGURE 2.14A
The Active Site of Chymotrypsin

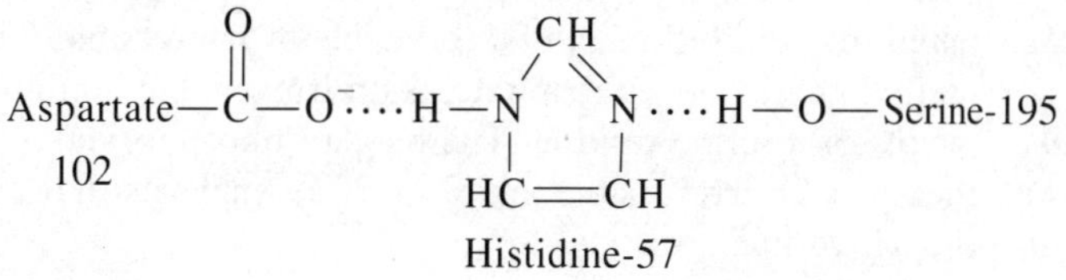

(a) Before the substrate binds

Research on chymotrypsin indicates that the active site contains a histidine residue, an aspartic acid residue and a serine residue at the active site. The groups work in concert to bring about the hydrolysis of peptide bonds.

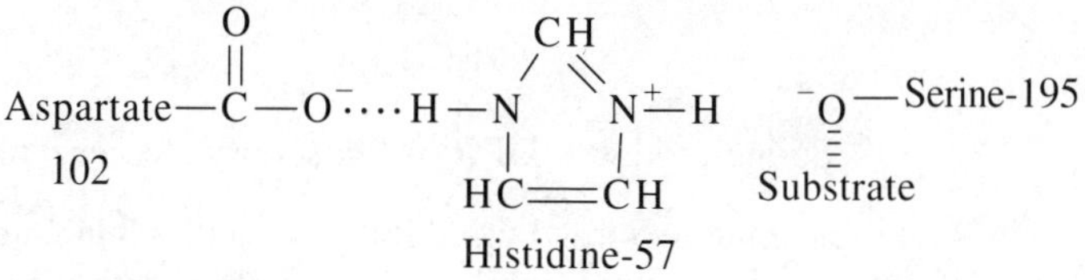

(b) After the substrate binds. Note that a proton has transferred from serine-195 to histidine-57, thus enabling serine to bind the substrate.

The three residues serine-195, histidine-57 and aspartic acid-102 work in concert. As a result, the serine residue acts as though it has an ionised —O⁻ group, which is something that it would not possess in aqueous solution at this pH. This illustrates a feature of enzyme catalysis. The enzyme brings together polar groups which interact to bring about reactions which would be inconceivable outside the realms of enzyme kinetics. An active site is a micro-environment which can alter the properties of chemical groups and enable them to do things which they cannot do in free aqueous solution.

CHECKPOINT 2.14

1. Explain the following terms: catalyst, enzyme, substrate, active site, specificity of enzyme action.

2. (*a*) How can the presence of a serine residue at the active site be demonstrated?

(*b*) How is a serine residue able to bond to a substrate?

3. The graph shows how the rate of an enzyme-catalysed reaction changes as the substrate concentration changes.

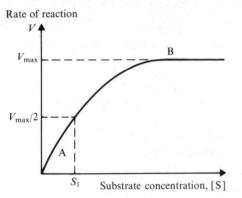

(*a*) What is the order of reaction (i) at A, (ii) at B?

(*b*) Why does the rate reach a maximum at high substrate concentration?

(*c*) What is the name given to V_{max} on the graph?

(*d*) What name is given to the concentration S_1?

2.15   ENZYME INHIBITION

Inhibitors decrease the catalytic activity of an enzyme. The inhibition may be reversible or irreversible.

The catalytic action of an enzyme can be decreased by substances called **inhibitors**. Inhibition can be **reversible** or **irreversible**. The reagent DFP, di-(1-methylethyl)-fluorophosphate, is an irreversible inhibitor of enzymes which act by means of a serine residue. The reagent phosphorylates serine residues and is therefore an irreversible inhibitor of all enzymes that employ a serine residue at the active site.

$$\text{Enzyme}\!-\!CH_2OH + F\!-\!\underset{\underset{\displaystyle OCH(CH_3)_2}{|}}{\overset{\overset{\displaystyle OCH(CH_3)_2}{|}}{P}}\!\!=\!\!O \longrightarrow \text{Enzyme}\!-\!CH_2O\!-\!\underset{\underset{\displaystyle OCH(CH_3)_2}{|}}{\overset{\overset{\displaystyle OCH(CH_3)_2}{|}}{P}}\!\!=\!\!O + HF$$

Serine residue + DFP → Phosphorylated enzyme + Hydrogen fluoride

Reversible inhibitors may be competitive or non-competitive.

Reversible inhibitors are divided into **competitive inhibitors** and **non-competitive inhibitors**. A competitive inhibitor can combine with the active site of the enzyme and compete with the substrate for binding at the active site. A competitive inhibitor resembles the substrate sufficiently to interact with the binding site.

FIGURE 2.15A
A Competitive Inhibitor

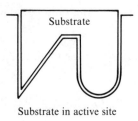

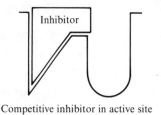

Substrate in active site

Competitive inhibitor in active site

Figure 2.15B shows the effect of a competitive inhibitor on the kinetics of an enzyme-catalysed reaction.

FIGURE 2.15B
Kinetics of Competitive
Inhibition

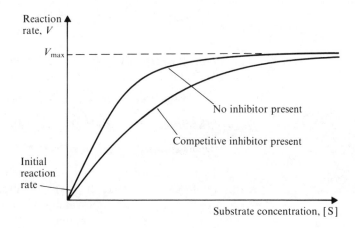

A competitive inhibitor bonds to the active site of the enzyme.

The extent of competitive inhibition depends on

- the concentration of the substrate
- the concentration of the inhibitor
- the relative strengths of the attraction of the active site for the substrate and for the inhibitor.

At high substrate concentration, it is possible to reach V_{max}, but the presence of the inhibitor increases the value of K_m.

A non-competitive inhibitor bonds to a different site and deforms the enzyme.

A non-competitive inhibitor can bind to a site on the enzyme other than the active site. As a result, the enzyme is deformed and cannot bind the substrate as effectively. A non-competitive inhibitor does not need to be of a shape similar to the substrate. The effect of a non-competitive inhibitor on the kinetics is shown in Figure 2.15C.

FIGURE 2.15C
Kinetics of Non-competitive Inhibition

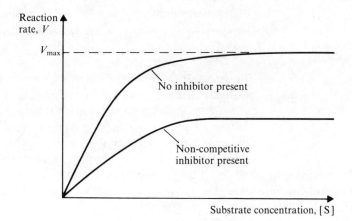

The extent of non-competitive inhibition depends on

- the concentration of the inhibitor
- the affinity of the enzyme for the inhibitor.

A non-competitive inhibitor may operate by combining reversibly with a group in the enzyme that maintains the three-dimensional structure of the enzyme molecule that is essential for catalytic activity. Heavy metal ions, e.g. Ag^+ and Hg^{2+}, inhibit enzymes by reacting with —SH groups.

$$\underset{CH_2SH}{\overset{H}{H_2N-\underset{|}{\overset{|}{C}}-CO_2H}} \text{ (aq)} + Ag^+ \text{(aq)} \longrightarrow \underset{CH_2SAg}{\overset{H}{H_2N-\underset{|}{\overset{|}{C}}-CO_2H}} \text{ (aq)} + H^+ \text{(aq)}$$

The enzyme glyceraldehyde phosphate dehydrogenase [see §9.2] has a sulphydryl group. Iodoethanoic acid reacts with this group to bind reversibly to the enzyme.

$$\text{Enzyme—SH} + ICH_2CO_2H \rightleftharpoons \text{Enzyme—S—}CH_2CO_2H + HI$$

Active enzyme   Iodoethanoic acid   Inactive enzyme

The enzyme is inactivated and the respiration of glucose cannot proceed.

A cell has hundreds of enzyme-catalysed reactions taking place in it. Many metabolic pathways involve a sequence of enzyme-catalysed reactions. The cell needs to be able to control these pathways, and switch substances from one pathway to another as its requirements change. Enzyme inhibition gives the cell a means of control. Consider a sequence,

$$A \xrightarrow{\text{enzyme X}} B \xrightarrow{\text{enzyme Y}} C \xrightarrow{\text{enzyme Z}} D$$

The product of an enzyme-catalysed reaction may inhibit the reaction. This situation is called feedback control.

If the concentration of product D becomes high, it may inhibit enzyme X non-competitively. When the product of a reaction inhibits its own formation, the situation is described as **feedback control**. The rate of the complete sequence is controlled by the concentration of the end product. You will find examples of feedback control in metabolic pathways, e.g. the cellular respiration of glucose, in §9.2.

2.15.1   INHIBITION OF ACETYLCHOLINESTERASE

Acetylcholine is a chemical transmitter of nerve stimuli. Nerve tissue contains the enzyme acetylcholinesterase which catalyses the hydrolysis of acetylcholine.

$$(CH_3)_3N^+CH_2CH_2OCOCH_3 + H_2O \longrightarrow (CH_3)_3N^+CH_2CH_2OH + CH_3CO_2H$$

Acetylcholine   Water   Choline   Ethanoic acid

The inhibition of acetylcholinesterase by curare is described.

The active site is thought to consist of two parts: an anionic part which binds the positively charged group of the substrate and the catalytic site which brings about the hydrolysis of the ethanoyl (acetyl) group. The sites are separated by a distance of 70 nm. The catalytic site is thought to consist of a hydroxyl group and an imidazole group. Acetylcholinesterase is competitively inhibited by curare. Curare molecules contain two quaternary ammonium groups separated by a distance of 140 nm. It has been proposed that each active site contains two negative sites about 140 nm apart so that both quaternary ammonium nitrogens can react with them.

FIGURE 2.15D
(a) Attachment of Biquaternary Ammonium Inhibitor, e.g. Curare
(b) Attachment of Two Acetylcholine (Substrate) Molecules

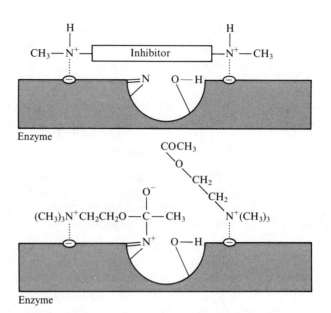

2.15.2 ANTIBIOTICS

Antibiotics are substances which kill or inhibit the growth of micro-organisms. Naturally occurring antibiotics are extracted from moulds, bacteria, yeasts, etc. Many of them are enzyme-inhibitors. Antibiotics are used to treat many diseases in humans and animals. Since animals have different enzymes from bacteria, only the bacteria are affected by antibiotics. They are used to speed the growth of poultry and livestock. The antibiotic 4-aminobenzenesulphonamide (trade name sulphanilamide) acts as a competitive inhibitor to prevent the growth of bacteria. Bacteria use 4-aminobenzoic acid in the synthesis of the vitamin folic acid. Since the structure of 4-aminobenzenesulphonamide is very similar to that of the substrate, 4-amino-benzenesulphonamide molecules compete for the active site on the bacterial enzyme and inhibit the synthesis of folic acid.

H_2N—⬡—CO_2H H_2N—⬡—SO_2NH_2

4-Aminobenzoic acid   4-Aminobenzenesulphonamide (sulphanilamide)

Many antibiotics are enzyme inhibitors. They inhibit bacterial enzymes without affecting the enzymes of animals.

Examples are sulphonamides, penicillins and cephalosporins.

The penicillins and cephalosporins are the most widely used antibiotics. They interfere with the process used by bacteria in constructing their cell walls. The structures of penicillins and cephalosporins are very similar to those of the compounds used by bacteria in building their cell walls. They compete for the active site on the enzyme which governs this process, critically weakening the cell walls and killing the bacteria.

2.16 COFACTORS

Amino acid side-chains are not effective in catalysing chemical reactions which require them to act as electron-acceptors. Some enzymes are able to catalyse reactions of this kind because they possess **cofactors**. These are non-protein substances of three types.

1. Activators, which bind to the enzyme transiently during the reaction. When metal ions serve as cofactors, they may

Proteins do not contain groups which enable them to catalyse reactions that require the catalyst to act as an electron acceptor. Cofactors are non-protein substances which assist enzymes in such reactions.

- act as electron acceptors at the catalytic site
- bind the substrate and enzyme together through the formation of a coordination complex
- stabilise the structure of the enzyme in its catalytically active shape.

The enzyme carbonic anhydrase requires one mole of Zn^{2+} ions per mole of enzyme. When zinc ions are removed by a complexing agent that will bind zinc more strongly, the enzyme becomes inactive. The enzyme amylase, which hydrolyses starch, requires the presence of chloride ion.

2. Prosthetic groups, which are organic compounds permanently bound to the enzyme, e.g. haem, which is the prosthetic group in the enzymes cytochrome (an electron carrier; see § 9.4) and catalase (which catalyses the decomposition of hydrogen peroxide). For the structure of haem see Figure 10.6.

3. Coenzymes are non-protein organic compounds which are not bound to the enzyme but are required for the functioning of the enzyme. Many coenzymes are derived from vitamins. Three coenzymes are formed from various vitamins of the B group. They act as hydrogen-carriers in enzyme-catalysed oxidation–reduction reactions, e.g. the oxidation of sugars. These coenzymes are NAD^+ (nicotinamide adenine dinucleotide), $NADP^+$ (nicotinamide adenine dinucleotide phosphate) and FAD (flavin adenine dinucleotide). The part of NAD^+ which acts as a hydrogen acceptor is shown attached to R, which represents the rest of the molecule. The reducing agent is shown as XH_2.

They may be activators, e.g. metal ions or prosthetic groups, e.g. haem or coenzymes, e.g. NADH⁺ and FAD. These coenzymes act as hydrogen-acceptors and electron acceptors in cellular respiration (§ 9.2).

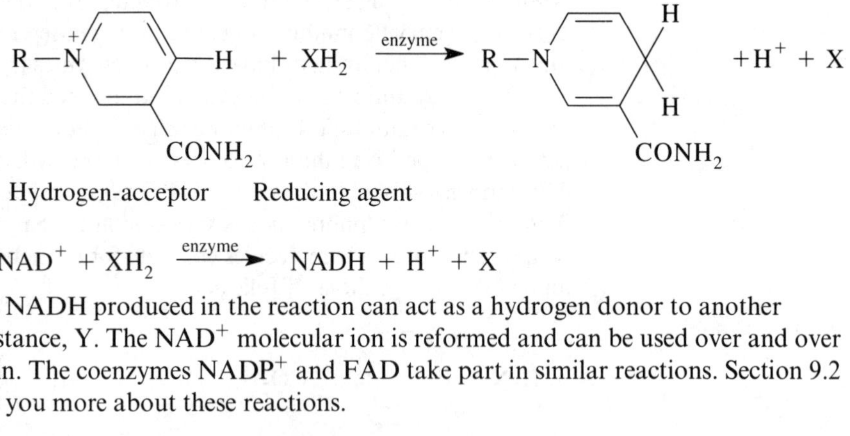

Hydrogen-acceptor Reducing agent

$$\text{or, } NAD^+ + XH_2 \xrightarrow{\text{enzyme}} NADH + H^+ + X$$

The NADH produced in the reaction can act as a hydrogen donor to another substance, Y. The NAD^+ molecular ion is reformed and can be used over and over again. The coenzymes $NADP^+$ and FAD take part in similar reactions. Section 9.2 tells you more about these reactions.

2.17 USES OF ENZYMES

Table 2.17A summarises some commercial and industrial uses of enzymes.

TABLE 2.17A Some Commercial and Industrial Uses of Enzymes

Application	*Enzymes*	*Uses*
Biological detergents	Proteases, produced from bacteria	Used in washing machines and for direct application.
Baking industry	Amylases	Catalyse hydrolysis of starch in flour to sugar. Yeast action on sugar produces carbon dioxide.
	Proteases	Reduce protein content of flour to make it suitable for biscuit making.

(*continued*)

Application	Enzymes	Uses
Dairy industry	Rennin, from calves' stomachs or from bacteria by genetic engineering	Used to hydrolyse protein in the manufacture of cheese.
	Lipases	Used to ripen blue mould cheeses, e.g. Danish blue, Rocquefort.
Brewing industry	Enzymes from barley are used in the mashing stage of beer production. Industrially produced enzymes are also used.	Convert starches and proteins into monosaccharides, amino acids and peptides. Yeast action on sugars produces alcohol.
Confectionery industry	Invertase	Incorporated in the solid centre of a chocolate, converts the solid centre into a liquid centre.
	Glucose isomerase as an immobilised enzyme	Used to make high-fructose syrups. Converts glucose, obtained from starch, into fructose, which is sweeter. The use of starch as a source of sweeteners is not permitted in the EC.
Paper industry	Amylases	Convert starch into a product of lower viscosity which is used for sizing and coating paper.
Rubber industry	Catalase	Liberates oxygen from hydrogen peroxide. This is used to make foam

Some of the many commercial and industrial uses of enzymes are tabulated.

CHECKPOINT 2.17

1. The rates of all enzyme-catalysed reactions increase as the substrate concentration is increased until they approach a maximum value. Which statement or statements explain this?

(*a*) All the substrate is used up.

(*b*) More and more enzyme molecules come into operation until all are combined with substrate.

(*c*) At high concentration, the substrate is an inhibitor of the enzyme.

2. (*a*) Explain what is meant by (i) an enzyme, (ii) a coenzyme.

(*b*) List the factors which affect the rate of enzyme reactions. Explain briefly why each of the factors changes the rate.

3. Explain the following terms, which relate to enzymes:

(*a*) active site

(*b*) lock and key theory

(*c*) competitive inhibitor

(*d*) optimum temperature

4. Explain the part played by the active site of an enzyme in (*a*) enzyme specificity (*b*) the catalytic action of enzymes and (*c*) competitive inhibition.

5. Explain how the lock and key comparison explains the mechanism of enzyme action. How does the induced fit theory change the lock and key theory?

2.18 IMMOBILISING ENZYMES

Batch processes use soluble enzymes and discard the enzyme after use.

Many industrial processes are **batch processes** which use soluble enzymes. The enzyme and substrate are mixed in a reaction vessel and left until a good yield of product has been formed. Usually the enzyme is discarded because recovery would be too costly. If the enzyme is expensive, only a small quantity may be used, and the reaction may be left to run for a long time with the possible formation of unwanted products from side-reactions. The cost of the enzyme used in an industrial process can be reduced if the enzyme can be recovered and used again. Making the enzyme insoluble by attaching it to an inert support achieves this aim. The technique is called **immobilising** the enzyme. The enzyme and its support can be packed into columns and substrate can be passed over the enzyme in a **continuous process** over long periods. Alternatively, the enzyme and its support may be in the form of insoluble granules which can be used in a batch reactor and then separated by centrifugation and re-used. Immobilising the enzyme allows a higher concentration of enzyme to be used, thus reducing the reaction time, reducing the formation of unwanted side-products and reducing operational costs.

2.18.1 METHODS OF IMMOBILISING ENZYMES

The technique of immobilising an enzyme attaches it to an inert support. This makes it possible to use the immobilised enzyme in a continuous process

Much work has been done on methods of immobilising enzymes. Binding the enzyme to a solid support may slightly alter the structure and therefore decrease the activity of the enzyme. The support material may be beneficial in reducing the ease with which the enzyme is denatured. It may also alter the optimal pH and temperature of the enzyme. As supports for the enzymes, insoluble polymers are used.

ADSORPTION

The enzyme may be immobilised by adsorption on to a resin . . .

The enzyme may be adsorbed on the support by means of interionic attractions between amino acid residues in the enzyme and charged groups on the support material, e.g. DAEA Sephadex [see Figure 2.18A(a)]. This kind of binding does not reduce the activity of the enzyme. The binding is weak, however, and some enzyme may be washed off its support.

TRAPPING

. . . or trapping inside a network of polymer molecules . . .

A solution containing the enzyme and a water-soluble monomer is prepared and the monomer is allowed to polymerise. The large enzyme molecules are trapped inside a network of polymer molecules, e.g. polyacrylamide gel [see Figure 2.18A(b)]. Small substrate molecules are able to penetrate the structure and interact with the enzyme. An enzyme immobilised in this way may be very stable, but it may lose some of its activity. The technique is inexpensive.

COVALENT BONDING

. . . or covalent bonding to a polymer support material.

One or more of the enzyme side-chains, e.g. $-NH_2$, $-OH$, $-CO_2H$, form covalent bonds with the polymeric support material, e.g. iodoethanoylcellulose [see Figure 2.18A(c)]. The bonding must allow the substrate to reach the active site. The production of a suitable support can be expensive.

2.18.2 EXAMPLES OF THE USE OF IMMOBILISED ENZYMES

HIGH-FRUCTOSE SYRUP

Sucrose, the sugar you find in the sugar bowl, is obtained from sugar beet and sugar cane. Starch is a more widespread carbohydrate. However, the hydrolysis of starch

FIGURE 2.18A
Immobilising an Enzyme
by (a) Adsorption,
(b) Trapping,
(c) Covalent Bonding

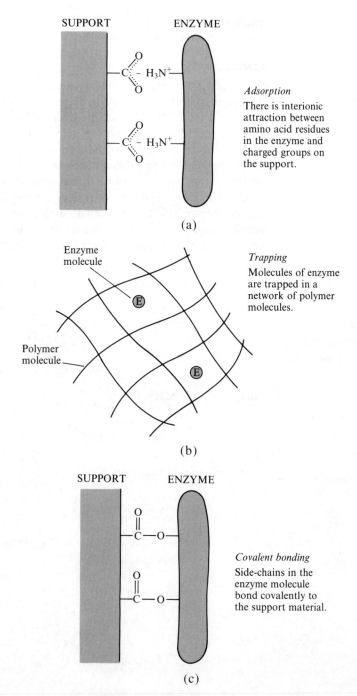

SUPPORT ENZYME

Adsorption
There is interionic attraction between amino acid residues in the enzyme and charged groups on the support.

(a)

Enzyme molecule

Trapping
Molecules of enzyme are trapped in a network of polymer molecules.

Polymer molecule

(b)

SUPPORT ENZYME

Covalent bonding
Side-chains in the enzyme molecule bond covalently to the support material.

(c)

Examples of the use of immobilised enzymes are ...

produces glucose which is less sweet than sucrose. It is an advantage to convert glucose into fructose, which is similar in sweetness to sucrose. Starch is partially hydrolysed and then treated firstly with glucoamylase to give glucose and secondly with glucose isomerase to convert glucose into fructose. The use of immobilised enzymes has halved the cost of production of high-fructose syrup [see § 3.3].

... the production of high-fructose syrup ...

RESOLUTION OF MIXTURES OF D- AND L-AMINO ACIDS

The naturally occurring L-amino acids are more useful than the D-forms. Synthesis always produces a racemic mixture, and it is necessary to separate the L-amino acid from this mixture. Aminoacylase catalyses the hydrolysis of the amide link in acyl-L-amino acids but is unable to hydrolyse the same bond in acyl-D-amino acids. The enzyme can therefore be used to isolate the L-amino acid from a racemic mixture. The

... the resolution of racemic mixtures of amino acids ...

unused D-isomer can be racemised and the separation repeated until all the mixture
has been converted into the L-form.

PENICILLIN PRODUCTION

Penicillin is a bactericide. It inhibits an enzyme which catalyses a stage in the
formation of bacterial cells walls. However, many strains of bacteria have become
resistant to penicillin. They secrete a penicillinase enzyme which hydrolyses an amide
group in penicillin and destroys its activity. It is possible to modify penicillin by
hydrolysing the amide group and then using the enzyme penicillin acylase to convert
the amine formed into a different amide. The second amide is not hydrolysed by the
enzyme in the resistant bacteria. It is called a **semi-synthetic penicillin**. In the industrial
production of semi-synthetic penicillins, immobilised penicillin acylase is used. The
enzyme is trapped in fibres of cellulose ethanoate or covalently linked to a modified
cellulose support. In the following reaction scheme, R is the part of the penicillin
molecule which is not involved in the reactions; R^1CO and R^2CO are acyl groups.

*. . . the production of semi-
synthetic penicillin.*

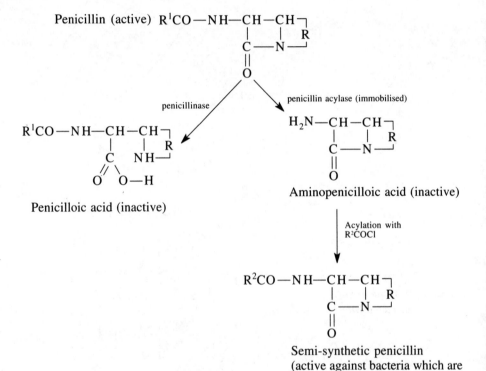

CHECKPOINT 2.18

1. When an enzyme-catalysed reaction is used in a batch
process, the enzyme is usually discarded after the reaction is
complete – or after a satisfactory yield of product has been
formed. Why is the enzyme not recovered for re-use?

2. Enzymes are often expensive. An economy can be made
by using a small amount of enzyme. What disadvantage does
this entail?

3. (*a*) What is an immobilised enzyme?

(*b*) What economic advantages are there to using
immobilised enzymes?

4. (*a*) Briefly describe two ways in which enzymes are
immobilised.

(*b*) Give two examples of the use of immobilised enzymes
and explain why they are chosen in preference to free
enzymes.

2.19 SOME NOVEL PROTEINS

2.19.1 ALGAE

The aim is to obtain protein for animal and human diets from single cell organisms which multiply rapidly. Algae which are rich in proteins have been harvested for food.

Algae are aquatic plants. They form the green covering that you see on many stagnant ponds. Many algae are rich in proteins, minerals and vitamins and are therefore a possible source of food. The idea is not new: centuries ago the natives of South America collected the alga *Spirulina* from the surface of lakes and dried it in the sun. They obtained a nourishing food with a mild taste at no expense. There is interest now in some hot countries, e.g. Israel, in cultivating algae in tanks and harvesting them for food. The investment required is only the cost of installing a simple tank and supplying fertiliser. The algae can be sun-dried. This work is still at the exploratory stage.

2.19.2 FUNGI

The cultivating of fungi which are rich in protein and harvesting them for food has had some commercial success.

Some fungi, e.g. mushrooms, are a welcome item in our diet. Since some fungi are rich in protein, the question of culturing them for food arises. The fungus *Fusarium* contains about 45% protein and has a high fibre content. The company Rank Hovis MacDougall developed a method of growing *Fusarium* on a wide variety of waste starches, e.g. wheat and potatoes. When the fungus is harvested and processed, it has a chewy texture similar to meat and a bland taste. It is made more palatable by the addition of flavourings and sold for human consumption. Protein obtained from fungi is called **mycoprotein**.

2.19.3 BACTERIA

Protein obtained from unicellular organisms is called **single-cell protein**. After a great deal of research, ICI succeeded in developing a method for growing bacteria as protein-rich foods. A genetically engineered bacterium named *Methylophilus methylotrophus* is grown on methanol which is a product of North Sea gas. The culture medium also includes mineral salts, ammonia and air. The bacteria are harvested and dried to yield a food which contains about 70% protein. The product is called Pruteen.

ICI has cultured the bacterium Methylophilus methylotrophus *to obtain a protein-rich animal feed, Pruteen.*

Pruteen is not suitable for the human diet because it contains too high a level of nucleic acids for some people to tolerate. It makes an excellent animal feed for chickens, calves, pigs, etc. The cost of manufacture is, however, too high to enable Pruteen to compete on the basis of price with other sources of animal feed, e.g. soya bean. Pruteen is not manufactured commercially, and the investment made by ICI in developing Pruteen has therefore not paid off directly. There has been a **spin-off** from the production of Pruteen as the new technology has found applications in the manufacture of other products in the firm.

QUESTIONS ON CHAPTER 2

1. The graph shows the results of experiments on the effect of increasing substrate concentration on the rate of an enzyme-catalysed reaction.

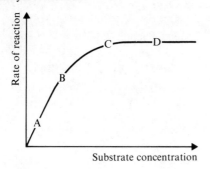

(*a*) What determines (i) the rate of the reaction between A and B, (ii) the shape of the graph between C and D?

(*b*) State two conditions which should be kept constant during the experiments.

(*c*) What has the experimenter measured in order to determine the rate of reaction?

(*d*) On a copy of the graph, draw the curve which you would expect in the presence of (i) a competitive inhibitor, (ii) a non-competitive inhibitor.

(*e*) Give an example of an enzyme, substrate and competitive inhibitor which could have been used in this study.

2. Define or explain the following terms, all of which relate to enzymes.

(*a*) cofactor, (*b*) activator, (*c*) coenzyme, (*d*) prosthetic group, (*e*) substrate, (*f*) active site, (*g*) isomerase

3. Describe the similarities and differences between the action of enzymes and inorganic catalysts.

4. (*a*) What is the *specific activity* of an enzyme?

(*b*) The enzyme pyruvate decarboxylase catalyses the reaction:

$$CH_3COCO_2^- + H^+ \rightarrow CH_3CHO + CO_2$$

Outline how you could measure the specific activity of the enzyme

(*c*) What is an *immobilised* enzyme?

(*d*) Describe one way that an enzyme could be immobilised.

(*e*) Suggest two advantages that an immobilised enzyme has over one that is not immobilised.

3

CARBOHYDRATES

3.1 FUNCTION

Carbohydrates include sugars, starches, glycogen and cellulose.

Sugar, starch and cellulose are **carbohydrates** [see *ALC*, § 31.18]. They consist of carbon, hydrogen and oxygen only and have the formula $C_m(H_2O)_n$. The simplest carbohydrates are the **sugars**, e.g. glucose, fructose and ribose. Sugars are **aldehydes** or **ketones** [see *ALC*, § 31.18]. **Monosaccharides** consist of a single polyhydroxyaldehyde or polyhydroxyketone. The most abundant monosaccharide is D-glucose, $C_6H_{12}O_6$. Most organisms derive their energy from the oxidation of glucose.

The monosaccharide sugar glucose is the source of energy released by oxidation in cellular respiration.

In all living organisms, carbohydrates perform the vital function of providing energy. When carbohydrates react with oxygen, carbon dioxide and water are formed and energy is released. Inside cells, this reaction takes place slowly so that energy is released in a controlled way. The process is called **cellular respiration** [see § 9.1]. Although most of our energy is obtained from the respiration of carbohydrates, living organisms also respire proteins and lipids.

Important polysaccharides are glycogen, starch and cellulose.

Carbohydrates act as the **food stores** glycogen in animals and starch in plants. Cellulose, the structural material of plants, is also a carbohydrate.

3.2 MONOSACCHARIDES

Monosaccharides are white, crystalline solids. They dissolve readily in water because they are able to form hydrogen bonds between their —OH groups and water molecules. They are insoluble in non-polar solvents.

Monosaccharides have the general formula $(CH_2O)_n$, where $n > 2$. Monosaccharides with the formula $C_6H_{12}O_6$ are **hexoses**. Monosaccharides with the formula $C_5H_{10}O_5$ are **pentoses**. Each molecule contains **one carbonyl group**. All the other carbon atoms are bonded to hydroxyl groups. The carbonyl group may be terminal, in which case the compound is an aldehyde sugar – an **aldose**. If the carbonyl group is not terminal, the compound is a ketone sugar – a **ketose**.

In the simplest case, the triose $(CH_2O)_3$, two isomeric formulae are possible:

Monosaccharides include hexoses of formula $C_6H_{12}O_6$ and pentoses of formula $C_5H_{10}O_5$. They include aldoses with an aldehyde group, —CHO, and ketoses with a ketone group, $\diagdown C{=}O$.

```
CHO                CH₂OH
|                  |
CHOH               CO
|                  |
CH₂OH              CH₂OH
```

2, 3-Dihydroxypropanal   1, 3-Dihydroxypropanone, a ketose
(glyceraldehyde), an aldose

On looking at the formula of glyceraldehyde more closely, you can see that it contains a chiral carbon atom [see *ALC*, § 5.1.4]. Two possible structures **A1** and **B1** can be drawn for glyceraldehyde [see Figures 3.2A and B].

FIGURE 3.2A
The Two Structures of
Glyceraldehyde

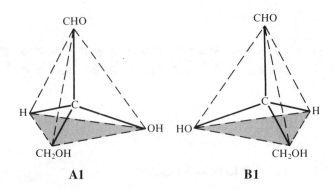

A1   **B1**

FIGURE 3.2B
The Enantiomers of
Glyceraldehyde

The structures **A1** and **B1** are **enantiomers** (mirror images). The two compounds with these structures have identical properties except that one rotates the plane of polarisation of plane-polarised light to the right and the other rotates it to the left. The formulae can be displayed as:

Monosaccharides exist as
D- and L-stereoisomers
called enantiomers.

$$
\begin{array}{cc}
\text{CHO} & \text{CHO} \\
| & | \\
\text{H}\!-\!\text{C}\!-\!\text{OH} & \text{HO}\!-\!\text{C}\!-\!\text{H} \\
\text{A2}\quad | & \text{B2}\quad | \\
\text{CH}_2\text{OH} & \text{CH}_2\text{OH}
\end{array}
$$

D-Glyceraldehyde   L-Glyceraldehyde

These formulae (**A2** and **B2**) which represent the structures shown as **A1** and **B1** are called **Fischer projections** after Emil Fischer, the famous sugar chemist. You will see the relationship if you make the models shown in Figures 3.2A and B. The isomer which has the formula with the OH group on the right-hand side when the formula is drawn as shown, with the CHO group at the top, is named D-glyceraldehyde. The isomer which has the formula with the OH group on the left of the formula as drawn in named L-glyceraldehyde. (Compare with amino acids, §2.2.)

If a molecule contains one chiral carbon atom, two enantiomers exist. The aldotetrose $CHO(CHOH)_2CH_2OH$ can have four configurations because it has two chiral carbon atoms. The four isomers fall into two pairs of enantiomers:

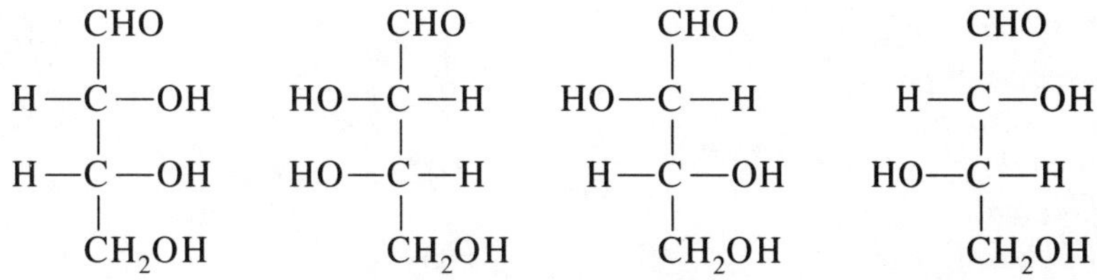

$$\begin{array}{cccc}
\text{CHO} & \text{CHO} & \text{CHO} & \text{CHO} \\
\text{H—C—OH} & \text{HO—C—H} & \text{HO—C—H} & \text{H—C—OH} \\
\text{H—C—OH} & \text{HO—C—H} & \text{H—C—OH} & \text{HO—C—H} \\
\text{CH}_2\text{OH} & \text{CH}_2\text{OH} & \text{CH}_2\text{OH} & \text{CH}_2\text{OH} \\
\text{D-Erythrose} & \text{L-Erythrose} & \text{D-Threose} & \text{L-Threose}
\end{array}$$

Diastereoisomers are stereoisomers which are not enantiomers.

The stereoisomers which are not enantiomers are called **diastereoisomers**. In this case, erythrose and threose are diastereoisomers. Both exist as D- and L-enantiomers. In the D-enantiomer the OH group on the bottom chiral carbon atom is on the right-hand side (as in glyceraldehyde). In the L-enantiomer, this OH group is on the left-hand side.

An aldohexose, e.g. glucose, has 16 isomers: 8 different compounds, each of which has D- and L-enantiomers.

In an aldohexose, $CHO(CHOH)_4CH_2OH$, there are four chiral carbon atoms. There are therefore 16 isomeric aldohexoses: 8 different compounds each of which has D- and L-enantiomers. The commonest aldohexose is D-glucose. A ketohexose, $CH_2OHCO(CHOH)_3CH_2OH$, has three chiral carbon atoms and therefore 8 isomers: 4 different compounds each with D- and L-enantiomers. The Fischer projections for three common sugars are shown below. Before Fischer began his work in 1886, only five naturally occurring hexoses were known. He synthesised and established the structures of 12 of the 16 aldohexoses.

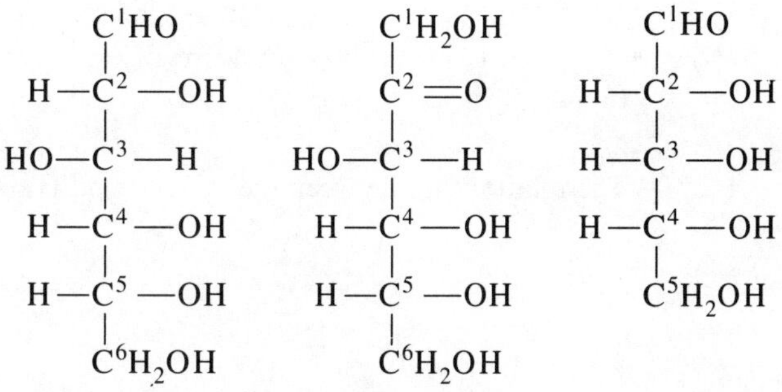

$$\begin{array}{ccc}
\text{C}^1\text{HO} & \text{C}^1\text{H}_2\text{OH} & \text{C}^1\text{HO} \\
\text{H—C}^2\text{—OH} & \text{C}^2\text{=O} & \text{H—C}^2\text{—OH} \\
\text{HO—C}^3\text{—H} & \text{HO—C}^3\text{—H} & \text{H—C}^3\text{—OH} \\
\text{H—C}^4\text{—OH} & \text{H—C}^4\text{—OH} & \text{H—C}^4\text{—OH} \\
\text{H—C}^5\text{—OH} & \text{H—C}^5\text{—OH} & \text{C}^5\text{H}_2\text{OH} \\
\text{C}^6\text{H}_2\text{OH} & \text{C}^6\text{H}_2\text{OH} & \\
\text{D-Glucose} & \text{D-Fructose} & \text{D-Ribose} \\
\text{(an aldohexose)} & \text{(a ketohexose)} & \text{(an aldopentose)}
\end{array}$$

The number of isomers is greater than predicted. To account for the number of isomers, the existence of ring structures is postulated.

The number of optically active centres in a monosaccharide is greater than that predicted from the formulae as shown above. For example, D-glucose appears in two stereoisomeric forms, α-D-glucose and β-D-glucose. To explain this, it is postulated that such molecules are six-membered rings formed by the addition of the —OH group at carbon-5 to the aldehyde group at carbon-1. The 6-membered ring structures are called **pyranose rings**. The formation of a ring has created another chiral carbon atom at carbon-1. The two isomers which differ only in the configuration of carbon-1 atom are **anomers**. Carbon-1 is called the **anomeric** carbon atom. The stereochemistry of carbon-1 is important because carbon-1 is involved in the polymerisation of sugars into polysaccharides. The configuration at carbon-1 can affect the shape of the polysaccharide chain and therefore its biological function.

They include the pyranose ring of glucose, the furanose ring of ribose and the pyranose and furanose rings of fructose. These are shown as Fischer projections or as Haworth projections.

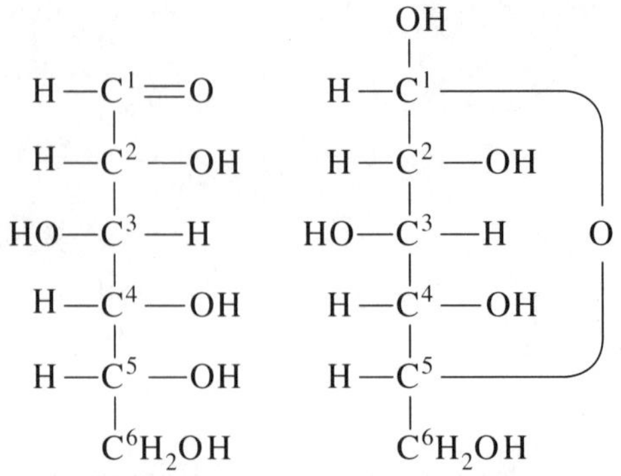

$$H-C^1=O$$
$$H-C^2-OH$$
$$HO-C^3-H$$
$$H-C^4-OH$$
$$H-C^5-OH$$
$$C^6H_2OH$$

D-Glucose

$$OH$$
$$H-C^1$$
$$H-C^2-OH$$
$$HO-C^3-H$$
$$H-C^4-OH$$
$$H-C^5$$
$$C^6H_2OH$$

O

D-Glucopyranose

A better way of representing the structure of D-glucopyranose is a **Haworth projection**.

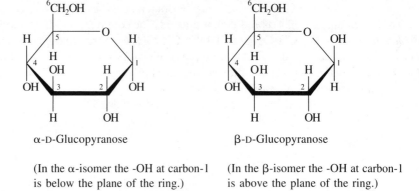

α-D-Glucopyranose

(In the α-isomer the -OH at carbon-1 is below the plane of the ring.)

β-D-Glucopyranose

(In the β-isomer the -OH at carbon-1 is above the plane of the ring.)

The relationship between the Fischer and Haworth projections is illustrated below.

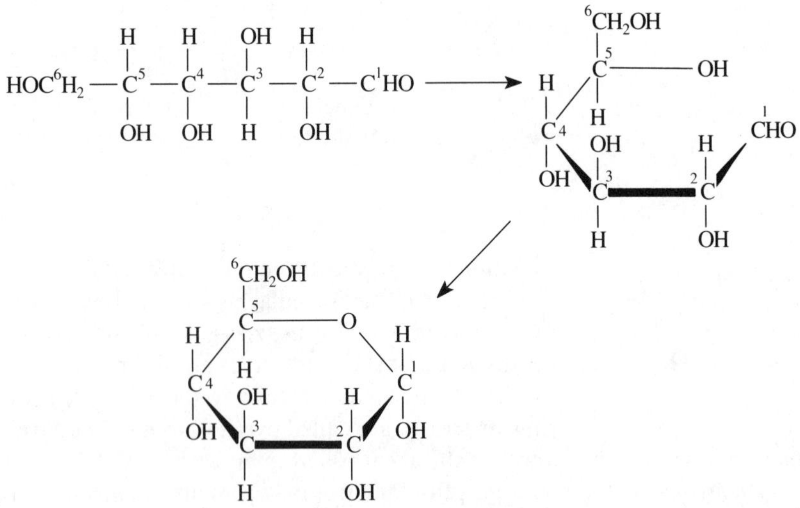

In fact the ring structure is not flat. The tetrahedral arrangement of bonds about the carbon atoms leads to a chair shape:

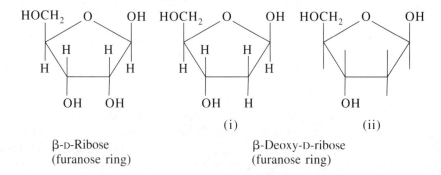

α-D-Glucopyranose

The pentose sugars form five-membered rings called furanose rings.

β-D-Ribose
(furanose ring)

(i)

β-Deoxy-D-ribose
(furanose ring)

(ii)

The five-membered ring is shown in Figure 3.2C. Frequently the H atoms in H—C—OH are omitted from a Haworth projection, as in formula (ii) for β-deoxy-D-ribose (see above).

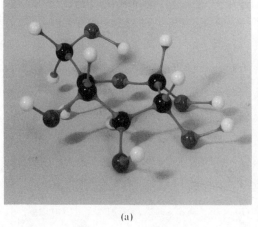

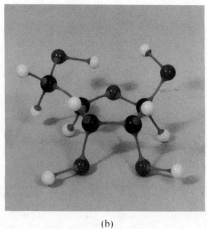

(a)

(b)

Fructose is a ketohexose and exists as the ring structures:

α-D-Fructose (pyranose ring)

α-D-Fructose (furanose ring)

In the free state, fructose exists in the pyranose form; in disaccharides and polysaccharides it exists in the furanose form.

3.3 DISACCHARIDES

Two monosaccharide molecules can combine by the elimination of a molecule of water between two —OH groups to form a **disaccharide**.

$$C_6H_{12}O_6 + C_6H_{12}O_6 \underset{\text{hydrolysis}}{\overset{\text{condensation}}{\rightleftharpoons}} C_{12}H_{22}O_{11} + H_2O$$

Monosaccharide + Monosaccharide $\rightleftharpoons$ Disaccharide + Water

Since water is eliminated, the reaction is described as a **condensation** reaction. The reverse is a **hydrolysis** reaction.

When starch is hydrolysed by a dilute acid, the disaccharide maltose is formed as an intermediate and is hydrolysed to glucose. This is the reaction which is used in the commercial production of glucose syrups. This reaction happens also during the sprouting of barley, and the products are important ingredients in beer production.

Disaccharides are formed when monosaccharides condense, that is, combine with the elimination of water. Monosaccharides are formed by the hydrolysis of disaccharides.

In maltose, two glucose molecules have condensed by the elimination of water to create a C—O—C linkage, which is called a **glycosidic linkage**. Since the linkage in maltose is between carbon-1 in one ring and carbon-4 in the other ring, it is called a 1,4-glycosidic linkage. Since it is in the α-position (below the plane of the rings as shown here), it is an α-1,4-glycosidic linkage.

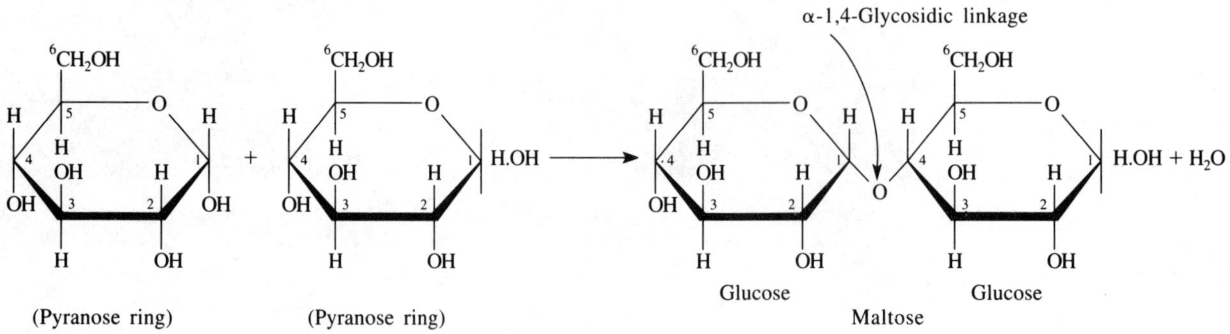

(Pyranose ring)   (Pyranose ring)   Glucose   Glucose
                                                                            Maltose

The —C—O—C— linkage which links sugar rings is called a glycosidic linkage. The α-1,4-glycosidic linkage of maltose is illustrated . . .

In maltose, the second pyranose ring contains an anomeric carbon atom, which corresponds to a —CHO group in the open-chain structure and can therefore be oxidised: maltose is a reducing sugar.

Lactose which occurs in milk (6–8% in human milk and 4–5% in cows' milk) is another disaccharide. Lactose has a free anomeric carbon atom and is therefore a reducing sugar.

...and the β-1,4-linkage of lactose...

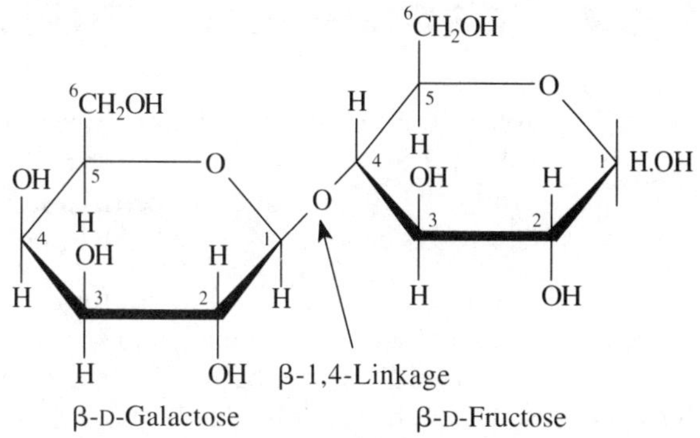

β-D-Galactose   β-1,4-Linkage   β-D-Fructose

Lactose

Sucrose (cane sugar) occurs in many fruits and vegetables. Sugar cane and sugar beet are used as sources of sucrose. Water is used to dissolve sucrose from the macerated plant, and the solution is crystallised to give granulated sugar. Sucrose is composed of α-D-glucose residues and β-D-fructose residues.

...and the α-1,2-linkage of sucrose.

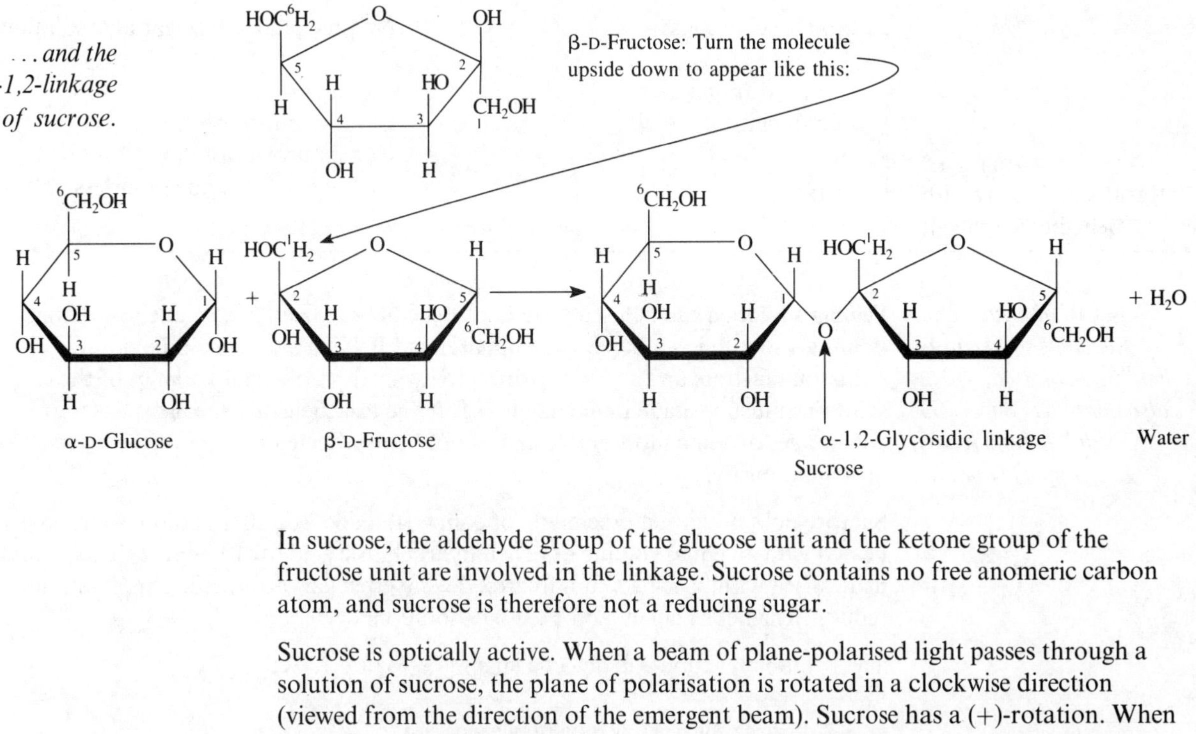

β-D-Fructose: Turn the molecule upside down to appear like this:

α-D-Glucose   β-D-Fructose   α-1,2-Glycosidic linkage   Water

Sucrose

In sucrose, the aldehyde group of the glucose unit and the ketone group of the fructose unit are involved in the linkage. Sucrose contains no free anomeric carbon atom, and sucrose is therefore not a reducing sugar.

Sucrose is optically active. When a beam of plane-polarised light passes through a solution of sucrose, the plane of polarisation is rotated in a clockwise direction (viewed from the direction of the emergent beam). Sucrose has a (+)-rotation. When sucrose is hydrolysed, either by warming with dilute hydrochloric acid, or by the enzyme sucrase, a mixture of glucose and fructose is formed. Glucose is weakly (+)-rotatary, and fructose is strongly (−)-rotatory.

$$(+)\text{-Sucrose} + \text{Water} \rightarrow (+)\text{-Glucose} + (−)\text{-Fructose}$$

Thus the (+)-rotation of sucrose is replaced by the (−)-rotation of fructose. The plane of polarisation of plane-polarised light is now rotated in the opposite direction (anticlockwise instead of clockwise). This change is called an **inversion**, the hydrolysis is called the inversion of sucrose, and the mixture of glucose and fructose is called 'invert sugar'.

Sucrose is optically active with a(+)-rotation. The reaction is called an inversion, and the mixture is called invert sugar.

The 'inversion' of sucrose is an important reaction in jam-making. When fruits are boiled with sucrose, acids in the fruits partially convert the sucrose into invert sugar. Invert sugar in the jam is less likely to crystallise than sucrose.

3.4 TESTING FOR REDUCING SUGARS

Apart from sucrose all the monosaccharides and disaccharides mentioned are reducing sugars. They contain a potential free carbonyl group (an aldehyde group or a ketone group). They can reduce copper(II) ions in boiling alkaline solution to copper(I) oxide. The reaction is used as a qualitative and quantitative test for reducing sugars.

Benedict's reagent is used to test for reducing sugars. It is a solution of copper(II) sulphate, sodium carbonate and sodium citrate. The reagent is warmed in a test tube immersed in a water bath of boiling water and an equal volume of an aqueous solution of a sugar is added. The mixture is kept in the warm water bath for several minutes. If the sugar is a reducing sugar, the blue colour of the solution disappears as copper(II) ions are reduced, and a reddish brown precipitate of copper(I) oxide appears. The time taken for the precipitate to appear varies from sugar to sugar. The colour depends on the concentration of reducing sugar [see Table 3.4A].

To test for a reducing sugar, Benedict's reagent can be used. It contains copper(II) sulphate which is reduced to red copper(I) oxide.

Concentration of reducing sugar	*Appearance of precipitate and solution*
No reducing sugar	No precipitate and clear blue solution
Concentration of reducing sugar increases.	Cloudy green solution Cloudy yellow solution Cloudy brown solution Red precipitate and colourless solution

TABLE 3.4A
Results of Testing with Benedict's Reagent

Fehling's solution can be used instead of Benedict's reagent. It is made by adding Fehling's solution A (copper (II) sulphate) and Fehling's solution B (sodium potassium tartrate and sodium hydroxide). A mixture of equal volumes of the two solutions must be made immediately before the test. Benedict's reagent has the advantages of being more stable and of not being affected by light or the presence of protein impurity.

An alternative to Benedict's reagent is Fehling's solution, which also contains copper(II) sulphate and forms copper(I) oxide.

Sucrose does not give a precipitate of copper(I) oxide. If a 10% sucrose solution is heated with an equal volume of bench hydrochloric acid for 10 minutes, cooled and neutralised with bench sodium hydroxide, 'invert sugar' is formed. This solution will reduce Benedict's reagent and Fehling's solution.

The Molisch reaction will detect a sugar [see Figure 3.4A].

FIGURE 3.4A
The Molisch Test for a Sugar. *Note*: Wear safety glasses and pour carefully

The Molisch reagent will detect a sugar by the formation of a violet colour with naphthalen-1-ol and concentrated sulphuric acid.

3 A dropping bottle of concentrated sulphuric acid
Concentrated sulphuric acid is poured **carefully** down the side of the test tube to form a layer at the bottom.

2 A few drops of sugar solution are added.

1 Naphthalen–1–ol
(enough to cover the bottom of the test tube) dissolved in a small volume of ethanol

4 A violet colour forms at the junction of the two layers.

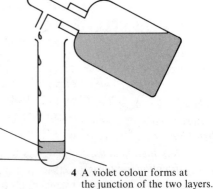

3.5 POLYSACCHARIDES

Polysaccharides have the general formula $(C_6H_{10}O_5)_n$. They include starch and glycogen.

Polysaccharides fulfil the important function of acting as **food stores**: starch in plants and glycogen in animals are polysaccharides.

Polysaccharides have high relative molecular masses. They are condensation polymers of monosaccharides of formula $(C_6H_{10}O_5)_n$. Polysaccharides are hydrolysed by acids and by enzymes to give monosaccharides. D-Glucose is the commonest monosaccharide unit in polysaccharides.

3.5.1 STARCH

Starch occurs as α-amylose and amylopectin. Both consist of D-glucose units joined by α-1,4-linkages and have $M_r = 5000$ to $500\,000$.

Starch occurs in two forms: α-amylose and amylopectin. Both consist of glucose units. α-Amylose consists of long unbranched chains of D-glucose units joined by α-1,4-linkages. The relative molecular mass varies from about 5000 to 500 000. When amylose is shaken with water, it does not form a true solution; it forms a colloidal suspension. The polysaccharide chains twist into helical coils, which are hydrated by water molecules and kept in suspension. Amylose gives a blue colour with iodine.

α-1,4-Glycosidic linkage   D-Glucose unit

α-Amylose

Amylose forms a colloidal suspension in water and gives a blue colour with iodine.

Amylopectin has highly branched molecules [see Figure 3.5A]. The length of each branch is 29–30 glucose units. The molecules consist of a chain of glucose units linked by α-1,4-linkages with side chains branching off through α-1,6-linkages. The molecules contain up to one million glucose units. Amylopectin forms colloidal solutions which give a reddish violet colour with iodine.

3.5.2 GLYCOGEN

Glycogen is similar to amylopectin, with a higher molar mass.

The very low solubility of starch and glycogen and their high molar masses are key factors in their function as storage materials.

Glycogen, the storage polysaccharide in animal cells, is similar to amylopectin. The molecules are more highly branched, at every 8 or 12 glucose units, through α-1,6-linkages. The relative molecular mass is much higher. Starch and glycogen are useful storage materials because they have very low solubility. In consequence, with their very low solubility and high molar masses, they exert very little osmotic pressure compared with the same mass of monosaccharide or disaccharide [see *ALC*, § 9.4]. They have little effect on the **osmotic pressure** of a cell (or, as sometimes expressed, the **osmotic potential** of a cell – the potential of the cell to draw water in by osmosis). When energy is required by the plant or animal, the polymers are easily hydrolysed by enzymes to give the monomers and dimers. The monosaccharides and disaccharides, being soluble, can be transported round the organism. On oxidation they release energy.

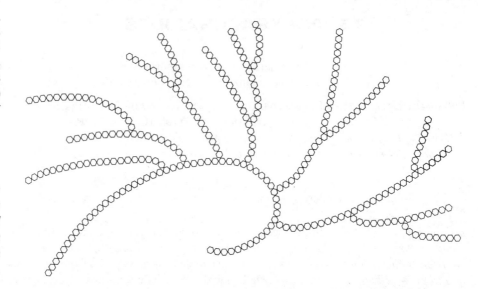

*Amylopectin has highly
branched molecules. It
forms a colloidal
suspension and gives a
reddish violet colour with
iodine.*

Amylopectin

*When energy is needed, the
polysaccharides are
hydrolysed to
monosaccharides and
disaccharides.*

When an aqueous starch solution is warmed at about $70\,^{\circ}\text{C}$ in a $1\ \text{mol}\ \text{dm}^{-3}$ solution of hydrochloric acid, it is hydrolysed. The progress of the reaction can be followed by testing for the presence of reducing sugars.

3.5.3   CELLULOSE

Cellulose is the structural component of plant cell walls. It is the major component of cotton and wood. Cellulose molecules consist of D-glucose units and have a relative molecular mass of about one million. The glucose units in cellulose are joined by β-glycosidic linkages between carbon-1 in the first glucose unit and carbon-2 in the second glucose unit and so on. In this diagram, the H atoms of the H—C—OH groups have been omitted. This is a common way of showing such projections.

β-1,4-Glycosidic linkage

Cellulose

In cellulose, the β-1,4-glycosidic linkage gives the molecules a linear conformation and they can therefore associate closely by hydrogen bonding. This structure gives cellulose its strength.

You see from the structural formula how every second glucose unit is flipped over so that the formation of β-1,4-glycosidic linkages creates an almost linear molecule with a rigid conformation. Making a model will help you to visualise the structure.

Several chains can therefore lie parallel to one another, and cellulose molecules form bundles of parallel chains. Hydrogen bonds form between chains and hold the bundles together. The alignment of chains and the bonding between them are responsible for the high mechanical strength of cellulose.

Cellulose is a food source for some animals, bacteria and fungi. The cellulase enzymes can hydrolyse the β-1,4-glycosidic linkage and therefore catalyse the digestion of cellulose to glucose. Cellulases are rare in nature, and most animals, including mammals, cannot utilise cellulose as food. There are mammals, however, e.g. cows, which do use grass and hay as food. Cows are called **ruminants**. In a part of their alimentary canal, called the **rumen**, they have bacteria which possess cellulase enzymes. After the cellulases have hydrolysed cellulose, ruminants can use the glucose produced as a source of energy. Although humans do not use cellulose as a food, it is an important component of the human diet because the fibre which it provides is the 'roughage' which gives bulk to food and helps it to pass through the alimentary canal.

Human enzymes cannot hydrolyse the β-1,4-linkage, and cannot therefore use cellulose as food. Bacteria in ruminants have cellulase enzymes which enable them to digest cellulose.

Commercially, cellulose is important in the manufacture of cellophane, of fabrics such as rayon, Dicel and Tricel and of paper.

QUESTIONS ON CHAPTER 3

1. Which of the following is or are (*a*) an aldose, (*b*) a ketose, (*c*) a triose, (*d*) a pentose, (*e*) a hexose?

A
$$CHO$$
$$H—C—OH$$
$$H—C—OH$$
$$H—C—OH$$
$$CH_2OH$$

B
$$CH_2OH$$
$$C=O$$
$$HO—C—H$$
$$H—C—OH$$
$$H—C—OH$$
$$CH_2OH$$

C
$$CHO$$
$$H—C—OH$$
$$CH_2OH$$

2. Draw the linear structures of (i) glucose and (ii) fructose.

3. Which of the following sugars give a positive test with Benedict's solution? (*a*) glucose, (*b*) fructose, (*c*) maltose, (*d*) sucrose, (*e*) ribose.

4. Angela and Adrian fancy a snack. Adrian eats a packet of potato crisps, but Angela, being on a diet, eats a stick of celery. Both food items consist of molecules which are polymers of glucose. Explain what difference between them made Angela choose the celery

5. (*a*) Describe two differences between a starch and cellulose.

(*b*) Why are humans unable to utilise cellulose as a dietary source of carbohydrate?

6. Maltose and sucrose are disaccharides with the formulae shown below. (You will notice that the H atoms in the H—C—OH groups are not shown in these projections, as explained in § 3.2.) Say which of them is a reducing sugar and which is a non-reducing sugar. Explain the reason for the difference.

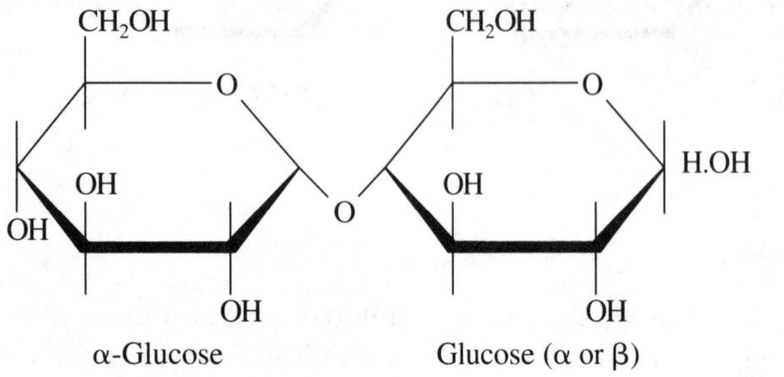

CH_2OH
CH_2OH
O
O
H.OH
OH
OH
O
OH
OH
OH
α-Glucose
Glucose (α or β)

Maltose

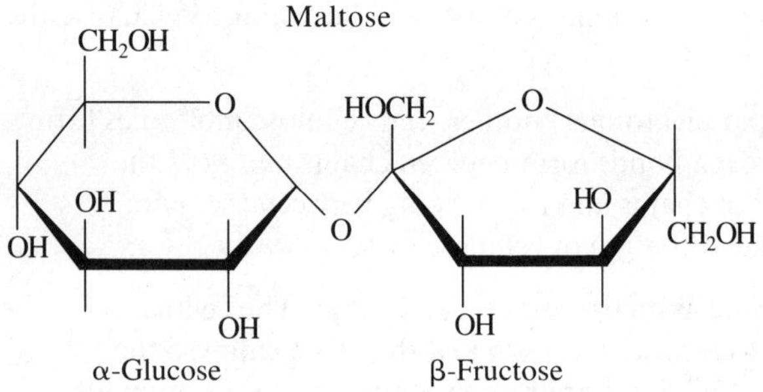

CH_2OH
O
$HOCH_2$
O
OH
HO
OH
O
CH_2OH
OH
OH
α-Glucose
β-Fructose

Sucrose

7. You are given three solutions labelled A, B and C. One contains starch, one contains glucose and one contains sucrose. Say what you could do to find out which solution is which.

8. (*a*) State two important sources of carbohydrates in our diet.

(*b*) What is the main function of carbohydrates in our diet?

(*c*) Name the monosaccharides of formulae:

CH_2OH
O
$HOCH_2$
O
OH
OH
HO
OH
O
CH_2OH
OH
OH

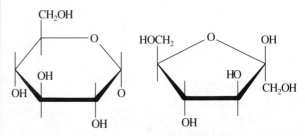

(*d*) Give the name and formula of the disaccharide formed from these two monosaccharides by means of an α-1,2-linkage.

4

LIPIDS

4.1 WHAT ARE LIPIDS?

Fats and **oils** and **waxes** are together called **lipids**. They are insoluble in water and have a number of important biological functions.

Lipids include fats, oils and waxes. Lipids are foods, protective layers, components of cell membranes, hormones and steroids.

- They are high-energy foods which can be stored for a long time in animals and plants.
- They form protective coatings, e.g. on plant leaves and round organs such as the kidneys in mammals.
- In mammals they form a layer under the skin which gives the body thermal insulation.
- They are the structural components of cell membranes.
- Some hormones are lipids; see § 4.10, 9.8.
- Steroids are lipids; see § 4.10.

Fats and oils can be extracted from biological material by crushing the material and mixing it with a non-polar solvent, such as trichloroethane or ethoxyethane. Lipids dissolve in the non-polar solvent.

Fats and oils are esters of glycerol with fatty acids; they are triacylglycerols or triglycerides.

Fats and oils are triacyl glycerols – esters of glycerol (propane-1,2,3-triol) with fatty acids (aliphatic acids with 16–22 carbon atoms). Waxes are esters of fatty acids with alcohols of high relative molecular mass. Glycerol, being a triol, can form a triester with a fatty acid such as octadecanoic acid (traditionally called stearic acid).

They may be simple glycerides, with three identical fatty acid groups ...

$$3\ C_{17}H_{35}\overset{O}{\underset{\|}{C}}{-}OH + \begin{matrix} HO{-}CH_2 \\ | \\ HO{-}CH \\ | \\ HO{-}CH_2 \end{matrix} \longrightarrow \begin{matrix} C_{17}H_{35}\overset{O}{\underset{\|}{C}}{-}O{-}CH_2 \\ | \\ C_{17}H_{35}\overset{O}{\underset{\|}{C}}{-}O{-}CH \\ | \\ C_{17}H_{35}\overset{O}{\underset{\|}{C}}{-}O{-}CH_2 \end{matrix} + 3H_2O$$

Octadecanoic acid + Propane-1,2,3-triol ⟶ Propane-1,2,3-triyl trioctadecanoate + Water
(stearic acid)   (Glycerol)   (Trioctadecanoylglycerol or tristearin, a fat)

... or mixed glycerides – with different fatty acid groups.

Water is eliminated by the reaction of —OH from each carboxyl group with —H from glycerol. An ester of glycerol is a **glyceride**; a triester is a **triglyceride**. If the three fatty acid groups are the same, as in the example above, the ester is a **simple glyceride**. If the fatty acids are different, as is the case in most naturally occurring fats and oils, the ester is a **mixed glyceride**.

The choice of which fat to use in a food often depends on the **melting temperature**. Fats used in confectionery, e.g. cocoa butter, must melt close to body temperature. For making flaky pastry, a fat with a high melting temperature is used so that layers of fat can be formed between sheets of dough. For cake-making, fats which melt over a wide temperature range should be used. Since fats consist of a mixture of triglycerides, they do not have sharp melting temperatures. To find the melting temperature, the capillary tube method shown in *ALC*, § 34.2, Figure 34.3 can be used. The temperature at which the fat clears and begins to run out of the capillary is called the **slip point**. Alternatively, a boiling tube containing the fat and a thermometer can be put into a beaker of water and the water heated until the fat melts. The melting temperature is the temperature at which all the fat has melted [see Figure 4.1A]. Butter starts to melt at about 35 °C and lard at over 50 °C.

Fats do not have sharp melting temperatures because they are not pure compounds, as shown in the figure. Fats are chosen for cooking purposes on the basis of the melting temperature.

FIGURE 4.1A
A Melting Temperature Curve for a Fat

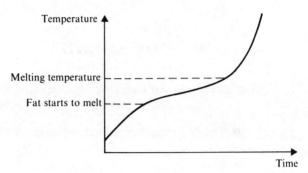

4.2 HOW TO TEST FOR LIPIDS IN FOODS

TEST 1: TRANSLUCENCY TEST

The food to be tested should be dried in an oven at 100 °C and tested as shown in Figure 4.2A.

FIGURE 4.2A
Testing Food for Lipids

The translucency test for lipids . . .

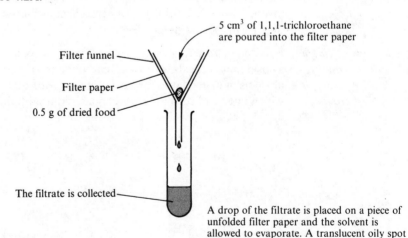

TEST 2: EMULSION TEST

A portion of the food (about 2 cm^3) is put into a test tube and ethanol (about 5 cm^3) is added. The tube is shaken to help lipids to dissolve. Then water (about 5 cm^3) is added, and the tube is shaken again. A cloudy white opalescent suspension shows the presence of a lipid. This happens because the addition of water to a solution of a lipid in ethanol results in an emulsion of tiny droplets of oil which reflect light and give an opalescent appearance.

. . . and the emulsion test for lipids.

4.3 HYDROLYSIS OF LIPIDS

Oils and fats develop an unpleasant smell if they are kept for too long. We describe it as a **rancid** smell. One cause of rancidity is the presence of **fatty acids** which have been formed through hydrolysis of the lipid by the water which it contains.

$$
\begin{array}{l}
CH_2O-\overset{\overset{\displaystyle O}{\|}}{C}-C_{17}H_{35} \\[4pt]
CHO-\overset{\overset{\displaystyle O}{\|}}{C}-C_{17}H_{35}\ (s)\ +\ H_2O(l) \longrightarrow C_{17}H_{35}CO_2H(aq)\ + \\[4pt]
CH_2O-\overset{\overset{\displaystyle O}{\|}}{C}-C_{17}H_{35}
\end{array}
\qquad
\begin{array}{l}
CH_2OH \\[4pt]
CHO-\overset{\overset{\displaystyle O}{\|}}{C}-C_{17}H_{35}\ (l) \\[4pt]
CH_2O-\overset{\overset{\displaystyle O}{\|}}{C}-C_{17}H_{35}
\end{array}
$$

Trioctadecanoylglycerol + Water ⟶ Octadecanoic acid + Dioctadecanoylglycerol
(tristearin)   (stearic acid)   (distearin)

Lipids develop a rancid smell in time. It may be caused by hydrolysis to form fatty acids with unpleasant smells. This is hydrolytic rancidity.

The hydrolysis takes place more rapidly in the presence of certain micro-organisms and of the enzyme lipase. Initially a diacylglycerol (diglyceride) is formed, but more molecules of fatty acid are liberated over a period of time. Some of the liberated fatty acids are volatile, and some have very unpleasant odours and flavours, which are described as 'rancid'. An example is butanoic acid which is formed when butter fat is hydrolysed. This type of rancidity is called **hydrolytic rancidity**.

The fatty acids have systematic names, e.g. octadecanoic acid (with 18 carbon atoms) and also traditional names, e.g. stearic acid.

The hydrolysis of lipids is not always a nuisance; it is put to good use. Glycerides are hydrolysed completely to glycerol and the salts of fatty acids by heating with alkali. The salts of fatty acids are used as **soaps**, and the hydrolysis is called **saponification**.

The hydrolysis of lipids gives the salts of fatty acids, which are used as soaps. The reaction is called saponification.

$$
\begin{array}{l}
C_{17}H_{35}\ \overset{\overset{\displaystyle O}{\|}}{C}-OCH_2 \\[4pt]
C_{17}H_{35}\ \overset{\overset{\displaystyle O}{\|}}{C}-OCH\ (s)\ +\ 3KOH(aq) \longrightarrow 3C_{17}H_{35}CO_2K(aq)\ + \\[4pt]
C_{17}H_{35}\ \overset{\overset{\displaystyle O}{\|}}{C}-OCH_2
\end{array}
\qquad
\begin{array}{l}
CH_2OH \\[4pt]
CHOH\ (l) \\[4pt]
CH_2OH
\end{array}
$$

Trioctadecanoylglycerol + Potassium ⟶ Potassium + Glycerol
(tristearin)   hydroxide   octadecanoate
                                          (potassium stearate)

4.4 SAPONIFICATION VALUE

The saponification value of a lipid is defined. A high saponification value indicates that the lipid has fatty acid groups of low relative molecular mass. A worked example illustrates this.

The **saponification value** of a lipid is the number of milligrams of potassium hydroxide needed to neutralise the fatty acids formed by the complete hydrolysis of one gram of the lipid. In the last example in §4.3,

Relative molecular masses are $M_r(\text{tristearin}) = 890$, $M_r(\text{KOH}) = 56$

Saponification value of tristearin is $(3 \times 56/890) \times 10^3 = 189$

The higher the relative molecular mass of the fatty acids, the lower is the saponification value of the lipid. A determination of the saponification value therefore gives some idea of the average relative molecular mass of the fatty acids in the lipid.

4.4.1 DETERMINATION OF SAPONIFICATION VALUE

About 1 g of fat is weighed accurately. A volume $50.0\,cm^3$ of $0.100\,mol\,dm^{-3}$ potassium hydroxide in ethanol is added, and the mixture is refluxed for 45 minutes or until the solution is clear. The solution is back-titrated against standard hydrochloric acid to find the volume of potassium hydroxide left and hence the volume of potassium hydroxide used in saponifying the fat. The mass in mg of KOH needed to saponify 1.00 g of fat can be calculated. This is the saponification value.

Worked example

Mass of fat $=$ 1.053 g
Volume of $0.100\,mol\,dm^{-3}$ potassium hydroxide added $=$ $50.0\,cm^3$
Volume of $0.100\,mol\,dm^{-3}$ hydrochloric acid need to neutralise excess
   KOH $=$ $22.0\,cm^3$
Volume of $0.100\,mol\,dm^{-3}$ potassium hydroxide used in
  saponification $=$ $28.0\,cm^3$
Mass of KOH used $=$ $28.0 \times 10^{-3} \times 0.100 \times 56 = 157\,mg$
Saponification value $=$ $157/1.053 = 149$

The **composition** of edible fats can be found by analysis. The fat is hydrolysed to glycerol and fatty acids. The fatty acids are converted into their methyl esters. These are separated by thin layer chromatography [see *ALC*, § 8.7.4]. They are invisible on the chromatogram until it is irradiated with ultraviolet light, when the different fatty acids can be seen. They can be identified by comparing the rate at which they move in thin layer chromatography with samples of methyl esters of known fatty acids.

CHECKPOINT 4.4

1. Explain how fats deteriorate on storage through hydrolysis.

2. You are asked to find the composition of a type of margarine. Describe how you would

(a) find its melting temperature

(b) test it for the presence of lipids

(c) hydrolyse a sample to obtain the fatty acids in it

(d) convert the fatty acids to their methyl esters

(e) separate the methyl esters by thin layer chromatography

3. A sample of fat weighing 0.840 g was refluxed with $25.0\,cm^3$ of an ethanolic solution of potassium hydroxide of concentration $0.125\,mol\,dm^{-3}$. The solution required $12.5\,cm^3$ of a $0.100\,mol\,dm^{-3}$ solution of hydrochloric acid for neutralisation. Calculate the saponification value of the fat.

4. (a) Write the structural formula for a mixed triacylglycerol formed between glycerol, hexadecanoic acid, octadecanoic acid and octadeca-9,12-dienoic acid.

(b) Is this compound likely to be solid or liquid at room temperature?

(c) Write the equation for the saponification of this triacylglycerol.

4.5 OXIDATION OF LIPIDS

Rancidity can result from the oxidation of lipids in air without the presence of enzymes. This is called autoxidation.

The oxidation of acylglycerols which occurs in air, without the presence of enzymes, is called **autoxidation**. The products are rancid. The oxidation is accelerated by the presence of transition metals, light and sources of free radicals. Autoxidation is a chain reaction with a free radical mechanism. [For free radicals, see *ALC*, § 26.3.7]. Among the products of autoxidation are **hydroperoxides**, ROOH. These have no taste or smell, but they decompose easily to form aldehydes, ketones and acids, which give oxidised fats and oils their rancid flavours. This type of rancidity is called **oxidative rancidity**.

Autoxidation is slowed down, but not eliminated, by the addition of **antioxidants**. Vitamin E, which is present naturally in many vegetable oils, is an antioxidant.

Antioxidants added to food include 2-hydroxypropane-1,2,3-tricarboxylic acid (citric acid), 2,6-di-(1,1-dimethylethyl)-4-methylphenol (BHT) and 2-(1,1-dimethylethyl)-3-methoxyphenol (BHA) [see § 13.1].

The products of autoxidation are aldehydes, ketones and acids with unpleasant smells. This is oxidative rancidity.

Citric acid   BHT   BHA

4.5.1   ACID VALUE

Autoxidation leads to the presence of acids in a lipid. The amount of acid which a lipid contains can be found by dissolving a weighed amount of the lipid in e.g. ethanol and titrating against a standard solution of an alkali. The **acid value of a lipid** is defined as the number of milligrams of potassium hydroxide needed to neutralise the acidity in one gram of the substance.

The acid value of a lipid is defined.

4.5.2   PEROXIDE VALUE

The amount of peroxide in a lipid is a measure of the extent to which oxygen has been taken up to form peroxides and is a measure of freshness. The amount of peroxide in a lipid can be found by analysis. An acidic solution of potassium iodide is added to a solution of the lipid in an organic solvent. The peroxide present oxidises iodide ion to iodine. The amount of iodine formed can be found by titration against a standard solution of sodium thiosulphate [see *ALC*, § 3.15.1].

$$ROOH + 2I^-(aq) + 2H^+(aq) \rightarrow I_2(aq) + ROH + H_2O(l)$$

$$I_2(aq) + 2S_2O_3{}^{2-}(aq) \rightarrow 2I^-(aq) + S_4O_6{}^{2-}(aq)$$

The peroxide value of a lipid is defined...

When 1 mole of oxygen adds to the lipid, 1 mole of hydroperoxide is formed. This liberates 1 mole of iodine which requires 2 moles of thiosulphate. The **peroxide value** of a lipid is defined as the amount in millimoles of oxygen, O_2, taken up per kilogram of oil. The legal limit is $10 \, \text{mmol kg}^{-1}$. A rancid taste develops when the peroxide value reaches $20–40 \, \text{mmol kg}^{-1}$.

Worked example   1.00 g of a fat was dissolved in a mixture of trichloroethane and glacial ethanoic acid. It was poured into $20 \, \text{cm}^3$ of 5% potassium iodide solution, and titrated against $2.00 \times 10^{-3} \, \text{mol dm}^{-3}$ thiosulphate, with starch as indicator. A blank titration was performed (without fat). The volume of thiosulphate used was $31.2 \, \text{cm}^3$; the blank was $1.2 \, \text{cm}^3$. Find the peroxide value of the fat and comment on it.

...and illustrated by a worked example.

Amount of thiosulphate $= (31.2 - 1.2) \times 10^{-3} \times 2.00 \times 10^{-3}$

$$= 60.00 \times 10^{-6} \, \text{mol}$$

Amount of iodine $= 30.00 \times 10^{-6} \, \text{mol}$

Amount of peroxide $= 30.00 \times 10^{-6} \, \text{mol} = 30.00 \times 10^{-3} \, \text{mmol}$

Mass of fat $= 1.00 \, \text{g} = 1.00 \times 10^{-3} \, \text{kg}$

Peroxide value $= 30.00 \times 10^{-3} / 1.00 \times 10^{-3} = 30 \, \text{mmol kg}^{-1}$.

This is much higher than the legal limit, and the fat will smell rancid.

4.6 FATTY ACIDS

Plants and micro-organisms synthesise fatty acids and acylglycerols from simple compounds. Animals synthesise some of their requirement of fatty acids and acylglycerols. Animals also ingest some plant and animal fats and oils and use them to synthesise the lipids which they require.

Fatty acids may be saturated or unsaturated. The unsaturated acids have a cis-configuration about the carbon–carbon double bond.

Fatty acids, e.g. stearic acid in the example given, are monocarboxylic acids with long, unbranched alkyl groups. Those which occur naturally have an even number of carbon atoms. They may be saturated or unsaturated. The unsaturated acids have the *cis*-configuration about the carbon–carbon double bond. In linoleic acid, there are double bonds between carbon 9 and carbon 10, and between carbon 12 and carbon 13. The configuration about both double bonds is *cis*; therefore the structural formula can be written:

Linoleic acid

Linoleic acid is an essential fatty acid; it cannot be synthesised in the body.

Linoleic acid is one of the two essential fatty acids in our diet since our bodies are unable to synthesise it. A deficiency can lead to scaly skin, loss of hair and a low growth rate. Its glyceryl esters are plentiful in sunflower oil [see Table 4.6A].

Fatty acid	M_r	*M.p./°C*	*Source*
Octadecanoic acid (stearic acid)	284	70	Lard, beef fat
Octadec-9-enoic acid (oleic acid)	282	16	Olive oil
Octadeca-9,12-dienoic acid (linoleic acid)	280	−5	Sunflower oil, soya bean oil, corn oil
Octadeca-9,12,15-trienoic acid (linolenic acid)	278	−11	Soya bean oil, linseed oil

TABLE 4.6A
Fatty Acids

The melting temperatures of fatty acids increase with relative molecular mass. They also depend on the degree of unsaturation: the number of carbon–carbon double bonds in the molecule. Table 4.6A shows how the melting temperatures of four acids with comparable relative molecular masses fall as the degree of unsaturation increases. When a fat is hydrolysed, the ratio of saturated fatty acids to unsaturated fatty acids is greater than that obtained from an oil. Even when fats are solid they contain a proportion of liquid acylglycerols. These give the fat a degree of plasticity; for example, some margarines can be spread when taken straight from the refrigerator.

Melting temperatures of fatty acids increase with relative molecular mass and decrease with the degree of unsaturation.

4.7 IODINE VALUE OF A LIPID

The carbon–carbon double bond in an unsaturated compound will add to iodine. The amount of iodine added is a measure of the degree of unsaturation of the compound.

$$R^1CH{=}CHR^2 + I_2 \rightarrow R^1CHICHIR^2$$

The iodine value of a lipid is defined.

The **iodine value** of a fat or an oil is the number of grams of iodine which react with 100 grams of the substance. In practice, the addition is carried out with iodine chloride, ICl, which is polar and therefore more reactive than iodine. For linoleic acid,

$$C_{17}H_{31}CO_2H + I_2 \rightarrow C_{17}H_{31}I_4CO_2H$$

The iodine value is $\dfrac{2 \times 254}{280} \times 100 = 181$

The iodine value of a mixture indicates the ratio of saturated lipids to unsaturated lipids in the mixture.

The iodine value of a saturated fatty acid is zero. The iodine value of a naturally occurring mixture of lipids gives an indication of the ratio of saturated lipids to unsaturated lipids in the mixture. Some iodine values are: coconut oil 8–10, butter 26–45, peanut oil 83–98, linseed oil 170–204.

4.7.1 HOW TO FIND THE IODINE VALUE OF A LIPID

Wij's reagent is prepared. A solution of a weighed mass of iodine (about 9 g) in 1,1,1,-trichloroethane is added to a solution of a weighed mass of iodine trichloride (about 8 g) in ethanoic acid and made up to $1.00\,dm^3$ with ethanoic acid. Some of the lipid (as oil or melted fat) is put into a weighing bottle and the required mass (about 1 g) is weighed out by difference into a conical flask. A little 1,1,1,-trichloroethane is added to dissolve the lipid, and $5.0\,cm^3$ of Wij's reagent is added. The flask is left to stand in the dark for 30 minutes. An excess of potassium iodide solution is added. The liberated iodine is titrated against standard sodium thiosulphate solution, using starch indicator near the end-point. A blank titration, without the lipid, is performed. The iodine value is calculated as follows.

A method of finding the iodine value is described.

If the volume of thiosulphate $= a\,cm^3$ of $0.10\,mol\,dm^{-3}$ and the blank $= b\,cm^3$ then the volume of thiosulphate used to titrate ICl $= (b - a)\,cm^3$ of $0.100\,mol\,dm^{-3}$. The equation for the reaction is

$$I_2(aq) + 2S_2O_3{}^{2-}(aq) \rightarrow 2I^-(aq) + S_4O_6{}^{2-}(aq)$$

Amount of thiosulphate $= (b - a) \times 10^{-3} \times 0.100 = 10^{-4}(b - a)\,mol$

Amount of $I_2 = \frac{1}{2}$ amount of thiosulphate

Mass of iodine $= 10^{-4}(b - a) \times 127$

Iodine value $=$ Mass of iodine per 100 g sample

$\qquad\qquad = (b - a) \times 1.27/(\text{Mass of sample})$

4.8 HYDROGENATION OF OILS

The melting temperature of an unsaturated oil can be raised by **hydrogenation** of the oil to form a solid fat. Usually, **partial hydrogenation** is employed, to give a fat which can be spread, e.g. soft margarine.

$$R^1CH{=}CHR^2 + H_2 \rightarrow R^1CH_2CH_2R^2$$

Hydrogenation is an important commercial reaction as there is more demand for solid margarines, spreads and shortenings than for cooking oils. Hydrogenation is carried out in the presence of finely divided nickel as catalyst. The progress of hydrogenation is followed by finding the iodine value of the product. It is stopped when the iodine value has fallen to the desired value. The reaction can be controlled by varying the pressure of hydrogen, the temperature, the type and concentration of the catalyst. In this way, partial hardening of oils to the desired extent can be achieved. The addition of hydrogen is accompanied by other reactions: the change of *cis*-unsaturated fatty acid groups into the *trans*-configuration and the migration of carbon–carbon double bonds along the carbon chains.

Hydrogenation and partial hydrogenation are used to turn unsaturated oils into solid fats. Finely divided nickel is used as a catalyst.

4.9 CALORIFIC VALUES OF LIPIDS

Fats and oils are rich energy sources. Their energy value is expressed in joules (J), in kilojoules (kJ) and in the older units, the calorie and kilocalorie. (4.18 kJ is the heat required to raise the temperature of one kg of water through $1\,^{\circ}C$, $4.18\,J = 1.00$ calorie.)

The **energy value** of a food is found by measuring the energy produced when one gram of the substance is completely oxidised to carbon dioxide and water. The combustion is carried out in a vessel called a **bomb calorimeter**. This can be filled with oxygen under pressure, ensuring that combustion is complete. When foods are oxidised in the body, the available energy is less than the heat of combustion. The difference is due to incomplete oxidation and incomplete absorption [see Table 4.9A].

The energy value of a food is found by combusting the food in pure oxygen in a bomb calorimeter.

Substance	Heat of combustion/$(kJ\,g^{-1})$	Available energy/$(kJ\,g^{-1})$
Fat	39	37
Carbohydrate	17	17
Protein	24	17

TABLE 4.9A
Average Energy Value of
Food Substances

4.10 STEROIDS AND TERPENES

Steroids and **terpenes** do not contain fatty acid residues, yet they are classified as lipids. Steroids are based on a structure composed of four rings:

Steroids and terpenes are classified as lipids.

The formulae of the steroids oestrone, testosterone and cholesterol are given opposite. Progesterone and cortisone are also steroids.

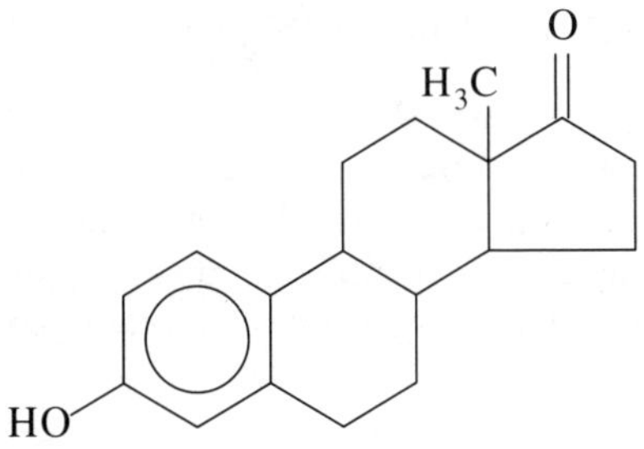

Oestrone (an oestrogen, a female sex hormone) 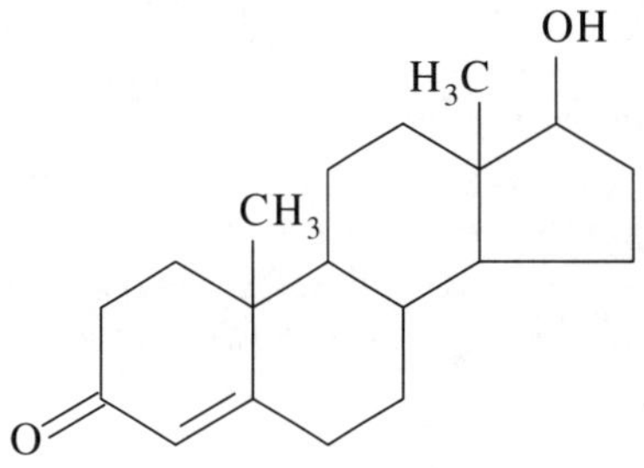   Testosterone (a male sex hormaone)

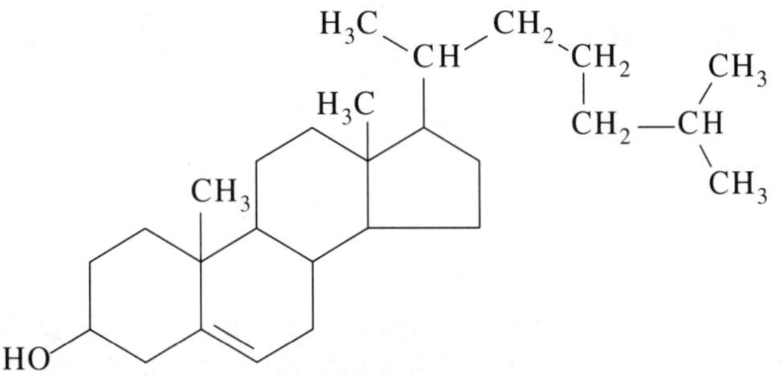

Cholesterol

Cholesterol is an important steroid, being a vital component of cell membranes and a starting point in the synthesis of other steroids.

Cholesterol is an important component of cell membranes. It is ingested in food and also synthesised in the liver. From cholesterol, animals synthesise other steroids, including sex hormones, certain vitamins, e.g. vitamin D, the bile acids, which assist in the digestion of lipids.

People suffer strokes and heart attacks when their blood vessels become narrow and restrict the flow of blood. The narrowing of blood vessels is often caused by the deposition of cholesterol on the inner walls. People with high levels of cholesterol in the blood are therefore prone to narrowing of the blood vessels. Diets which are high in cholesterol cause a rise in blood cholesterol levels, and there has been a big drive to get people to cut down the amount of cholesterol in their diet. However, the question is more complicated than this as other factors, including blood pressure, exercise, age, sex and heredity come into the picture. The body is able to manufacture cholesterol, and the amount which it manufactures is governed by a person's genes.

A diet that is high in cholesterol is suspected of promoting strokes and heart attacks.

Some studies have shown that eating food containing a high proportion of saturated fats raises the level of cholesterol in the blood. The **P/S ratio** is given by:

P/S ratio = Mass of polyunsaturated fat in diet/Mass of saturated fat in diet

In 1984 the P/S ratio for the average UK diet was 0.25, and a Government report suggested that a P/S ratio of 0.45 would reduce the risk of heart disease. The cholesterol picture is complicated, however, as you will see from the list of essential functions of cholesterol given above, and research is continuing.

Steroids similar to testosterone increase the rate of protein synthesis in both men and women and are called **anabolic steroids**. They are famed for their ability to increase the size of muscles and have been used by some athletes to improve their performance. These steroids have unwanted side-effects, such as kidney damage, development of secondary male sexual characteristics in women and impotence in men. In the interests of safety and fair play, the use of steroids in sport has been banned.

Anabolic steroids promote muscle development. Their use in sport has been banned.

Terpenes are lipids. They are responsible for the scents and flavours in 'essential oils' of plants, e.g. camphor and menthol in mint. They include:

- Gibberellins, which are plant growth substances
- Phytol, which is present as a phytyl group in chlorophyll [see Figure 10.6A]

● Carotenoids, which are pigments [see Figure 10.6A]
● Natural rubber

Terpenes are lipids which include essential oils of plants and plant growth substances, e.g. gibberellins.

The fungus *Gibberella* was studied because it causes disease in rice seedlings. Infected seedlings become exceptionally tall and spindly and eventually die. The active substances in the fungus were identified and named **gibberellins**. There are about fifty of them – all terpenes. The structure of one of them, gibberellic acid, is:

Gibberellic acid

Gibberellins are most abundant in young plants. They are **plant growth substances**. In the right concentration, they promote growth without causing the plant to overgrow its strength as the rice seedlings did. Gibberellins are used commercially in controlled quantities for several purposes:

● to promote the growth of stems
● to make dormant seeds germinate
● to increase crops of pears, tangerines, etc.
● to reduce losses in apple crops through frost damage
● to reduce losses in sugar cane crops through low temperatures

CHECKPOINT 4.10

1. Explain how fats deteriorate on storage through oxidation.

2. Describe how to test a fat to find out whether it contains peroxides.

3. (a) What is meant by the statement that peanut oil has an iodine value of 85?

(b) Describe how you would find the iodine value of a lipid.

(c) What is the importance of this information?

4. A given mass of margarine reacts with 2.5×10^{-3} mol iodine and requires 4.5×10^{-3} mol sodium hydroxide for complete hydrolysis. Calculate the average degree of unsaturation in the mixture of triacylglycerols (the average number of double bonds per molecule).

5. A 0.10 g sample of a vegetable oil was dissolved in a solvent and then added to 10.0 cm^3 of Wij's solution. The mixture was allowed to stand in the dark for 30 minutes. After 15 cm^3 of 10% potassium iodide were added, the liberated iodine was found to react with 5.8 cm^3 of 0.100 mol dm^{-3} sodium thiosulphate solution. A blank titration (without the oil) required 12.5 cm^3 of the same thiosulphate solution.

(a) State what is meant by the *iodine value* of an oil.

(b) What is Wij's solution?

(c) Write an equation for the reaction between thiosulphate and iodine.

(d) What is the reaction that occurs when the mixture of oil and Wij's solution is left to stand?

(e) What is the difference between the volumes of thiosulphate solution used in the blank and with the oil?

What amount of thiosulphate does this volume of solution contain?

What amount of iodine reacts with this amount of thiosulphate?

What mass of iodine is this? ($A_r(I) = 127$)

Calculate the iodine value of the oil.

6. (a) Explain why some athletes use anabolic hormones.

(b) What led the Olympic Committee to outlaw this practice?

7. Explain the apparent contradiction in employing gibberellins, which kill young plants, to increase commercial crops.

4.11 PHOSPHOGLYCERIDES

Phosphoglycerides (or phospholipids) are a major component of cell membranes. In these lipids, one of the hydroxyl groups of glycerol is esterified with an ester of phosphoric acid and the other two are esterified with fatty acids. The phosphate group is ionised. This gives rise to behaviour different from that of the unionised lipids that we have met so far.

$$
\begin{array}{c}
\text{CH}_2 - \text{O} - \overset{\displaystyle O}{\overset{\|}{\text{C}}} \\
\text{R}^2 - \text{C} - \text{O} - \overset{|}{\text{C}} - \text{H} \qquad \text{O} \; \text{R}^1 \\
\overset{\|}{\text{O}} \qquad \text{CH}_2 - \text{O} - \overset{\|}{\text{P}} - \text{O} - \text{R}^3 \\
\underset{\text{O}^-}{|}
\end{array}
$$

The groups R^1, R^2 and R^3 are alkyl groups. The phosphate ester group is a polar group, and the fatty acid ester groups are non-polar groups; therefore the phosphoglyceride can be described in terms of a polar head and non-polar tails.

Phosphoglycerides or phospholipids are major components of cell membranes. The three hydroxyl groups of glycerol are esterified with two fatty acids and phosphoric acid. The negative charge on the phosphate group gives rise to properties different from the lipids studied so far.

$$
\left.
\begin{array}{c}
\text{R}^3 \\
| \\
\text{O} \\
| \\
\text{O} = \text{P} - \text{O}^- \\
| \\
\text{O}
\end{array}
\right\} \text{Polar head}
$$

$$
\left.
\begin{array}{c}
\text{CH}_2 - \text{CH} - \text{CH}_2 \\
| \qquad | \\
\text{O} \qquad \text{O} \\
| \qquad | \\
\text{C} = \text{O} \quad \text{C} = \text{O} \\
| \qquad | \\
\text{CH}_2 \quad \text{CH}_2 \\
| \qquad | \\
\text{CH}_2 \quad \text{CH}_2 \\
| \qquad | \\
\text{R}^1 \qquad \text{R}^2
\end{array}
\right\} \text{Non-polar tails}
$$

Molecules of phosphoglycerides have hydrophilic groups and hydrophobic groups, which make the molecules group together to form micelles and to form bilayers.

Phosphoglycerides are waxy solids. They dissolve very slightly in water to give a true solution. Molecules of phosphoglycerides group together to form **micelles**, in which the hydrophilic polar heads are solvated by water molecules and the hydrophobic non-polar tails are shielded from water molecules. Molecules also form **bimolecular layers (bilayers)** in which the hydrophilic polar 'heads' lie on the outside of the layer so they can be solvated by water molecules and the hydrophobic hydrocarbon 'tails' are shielded from water molecules.

FIGURE 4.11A
Phosphoglycerides
(Phospholipids)

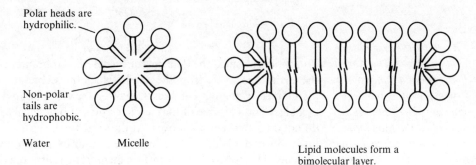

Polar heads are hydrophilic.

Non-polar tails are hydrophobic.

Water   Micelle

Lipid molecules form a bimolecular layer.

4.12 CELL MEMBRANES

*Cell membranes contain
lipids, which are chiefly
phospholipids, and
proteins.*

Cell membranes contain lipids and proteins. The proportions vary with the type of membrane and with the organism, but a common ratio is 40% lipid to 60% protein. The lipids are mainly phosphoglycerides, which have polar groups. The lipids and proteins in a membrane are associated by van der Waals interactions between their non-polar regions.

The surface membrane of a cell [Figures 1.1A and B] has three main functions to perform:

1. It has to keep the cell intact by means of its strength, insolubility and protective nature.

*A cell surface membrane
has the functions of
keeping the cell intact,
recognising and bonding to
hormones and
neurotransmitters, and
selecting which substances
can pass into and out of the
cell.*

2. It has to provide a surface which will recognise and bond to hormones, neurotransmitters (released from nerve endings) and other cells of the same tissue.

3. It has to act as a selective barrier to substances in the cell's environment, allowing some to pass and excluding others. To do this the surface must be able to discriminate between harmful substances that would damage the cell and essential substances required by the cell. The surface membrane maintains the correct ion concentration inside the cell so that pH, osmotic pressure and the electric potential of the cell remain constant. Conditions outside the cell may vary widely, and it is the cell surface membrane that maintains the balance inside the cell.

Owing to the importance of the cell surface membrane, research into the way in which it functions is the subject of intense activity, and theories are constantly being modified.

In the cell surface membrane, phospholipid molecules organise themselves into a double layer, with the non-polar tails on the inside and the polar heads on the outside, where they can readily bond to proteins and glycoproteins (compounds of proteins and sugars) [see Figure 4.12A]. The current theory of cell surface membrane structure is a core of phospholipid molecules arranged in a bilayer with its surface covered with protein molecules, some of which penetrate right through the bilayer [see Figure 4.12A].

FIGURE 4.12A
The Fluid Mosaic Model
of a Cell Surface
Membrane
(simplified picture)

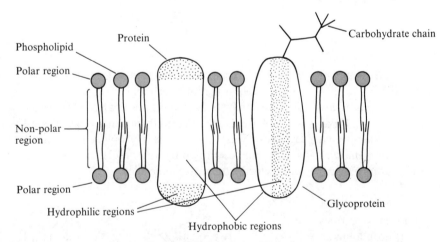

*A cell surface membrane
consists of a bilayer of
phospholipid molecules in
which protein molecules
are embedded. This model
is called the fluid mosaic
model of membrane
structure.*

This model is called the **fluid mosaic model** of membrane structure. The **fluid nature** of the membrane comes from the way in which the secondary and tertiary structure is maintained by many weak bonds, making the phospholipid bilayer very fluid and flexible. Fluidity is one of the characteristics of biological membranes: cell membranes do not crack when a cell moves. At the same time, because of the large number of internal van der Waals attractions, the membrane is also strong and resistant to solvents. This is how the cell surface membrane is able to carry out its function of

Proteins in the membrane, integral proteins, provide channels through which substances can move. Peripheral proteins are attached to the surface.

keeping the cell intact. Under a microscope, the two rows of phosphate heads show as two dark bands and the hydrocarbon chains as a light band in between them.

The mosaic nature of the membrane is the **mosaic of proteins** floating in the phospholipid fluid and embedded in its surface. The proteins in the membrane are both fibrous and globular. Some are embedded in the bilayer in between phosphoglycerides (phospholipids). These are described as **integral proteins**. Many integral proteins span from one surface of the membrane to the other. They provide channels through which substances can move [see Figure 4.12B]. other proteins are attached to the surface of the membrane. These are the **peripheral proteins** [see Figure 4.12C].

The membrane proteins take part in cellular processes of three kinds:

The proteins in the membrane are:

transport proteins, involved in the passage of substances through the membrane

catalytic proteins, which are enzymes

receptor proteins, which bind hormones etc.

● **Transport proteins** are involved in the passage of substances into and out of the cells.

● **Catalytic proteins** are enzymes, catalysing the reactions which occur at the membrane [see § 2.7 for enzymes].

● **Receptor proteins** bind another substance, e.g. a hormone, on the outer surface of the membrane and provide a signal that affects cell reactions [see below].

The second function of the cell surface membrane is to recognise and bond to substances that are important in the metabolism of the cell, e.g. hormones, and to recognise and bond to other cells. This function is thought to depend on **receptor proteins**. They are usually glycoprotein molecules embedded in the external surface of the membrane [see Figures 4.12A and C]. Glycoprotein molecules have polysaccharide chains attached to polypeptide chains. With the polypeptide chains embedded in the phospholipid membrane and the polysaccharide chains pointing outwards into the medium outside the cell, these polysaccharide chains are well placed to bond to complementary polysaccharide chains on the surfaces of other cells. In this way cells can assemble to form tissues and organs. If the polysaccharide chains of their glycoproteins are not complementary, cells will not stick together; they will move on until they recognise another cell of the correct type. This function of the cell surface membrane is particularly important when the organism is developing and growing fast and many new cells are being produced and incorporated in developing organs.

Receptor proteins are usually glycoproteins – compounds of polysaccharides and proteins. Their polypeptide chains are embedded in the membrane and their polysaccharide chains are free to bond to polysaccharide chains on the surfaces of other cells. This is what happens when cells assemble to form tissues.

The third function of the cell surface membrane is to regulate the passage of substances into and out of the cell. Membranes can select the materials which move through them: they are **selectively permeable**. If a membrane consisted entirely of a phospholipid bilayer, it would be impermeable to anything that was not lipid-soluble. However, the mosaic of proteins solves this problem by providing channels through which water and water-soluble substances can move [see Figure 4.12B]. Figure 4.12C shows the structure of phospholipids, globular and cellular proteins and glycoproteins in a cell surface membrane.

FIGURE 4.12B
A Channel Protein

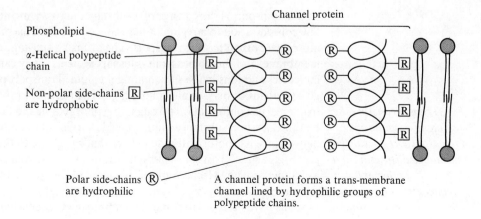

Phospholipid

α-Helical polypeptide chain

Non-polar side-chains R are hydrophobic

Channel protein

Polar side-chains R are hydrophilic

A channel protein forms a trans-membrane channel lined by hydrophilic groups of polypeptide chains.

FIGURE 4.12C
The Fluid Mosaic Model
of a Cell Surface
Membrane

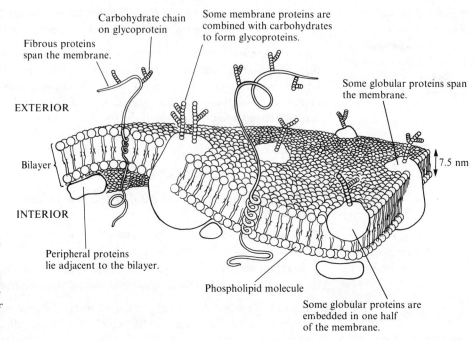

*Cell surface membranes
are selectively permeable.
The proteins embedded in
the membrane provide
channels for the passage of
water-soluble substances.*

Transport of materials across membranes may occur by one of two methods, **passive
transport** and **active transport**.

4.12.1 PASSIVE TRANSPORT

*Transport across the
membrane may be by
passive transport, in which
a substance diffuses from a
region of higher
concentration to lower
concentration by simple
diffusion or by facilitated
diffusion.*

A substance may diffuse across a membrane from a region of high concentration to a
region of low concentration until the concentrations are the same and diffusion
ceases. This is called **passive transport**. It may occur by simple diffusion or by
facilitated diffusion.

● **Simple diffusion** allows substances which are soluble in lipids to diffuse through the
phospholipid bilayers. Water and dissolved ions cannot do this; they diffuse through
channel protein molecules which span the membrane to form pores [see Figure 4.12B].

● **Facilitated diffusion** involves a carrier. The carrier molecule, associated with the
diffusing molecule or ion, travels from one side of the membrane to the other.

4.12.2 ACTIVE TRANSPORT

*Transport across the
membrane may be by
active transport, in which a
substance diffuses from a
region of lower
concentration to higher
concentration.*

Active transport is the passage of a substance across a membrane from a region of
low concentration to a region of high concentration. This process consumes much
more energy than passive transport. The diffusing substances require help to cross the
membrane, and this comes from **transport proteins**. They can recognise and bond to
molecules and ions in the surrounding medium. Transport proteins may be
glycoproteins or may be phosphorylated. The transport protein molecule may be able
to change its conformation through phosphorylation and dephosphorylation. After
binding a certain substance on the outside surface of the cell, the transport protein
may change its conformation to deliver that substance to the interior of the cell
[see Figure 4.12D]. In some cases, the transport protein may pass the bound species to
a channel protein. Mechanisms of this sort are vital in the transport of many
substances across the membrane. A number of transport proteins and channel
proteins have been isolated and their amino acid sequences mapped.

All the functions of the cell surface membrane are made possible by the fluid lipid environment which does not restrict either lateral or transverse movement.

FIGURE 4.12D
Assisted Transport Across
a Membrane

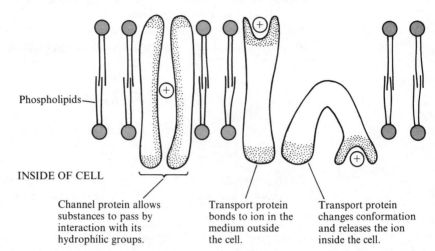

OUTSIDE OF CELL

Phospholipids—

INSIDE OF CELL

Channel protein allows substances to pass by interaction with its hydrophilic groups.

Transport protein bonds to ion in the medium outside the cell.

Transport protein changes conformation and releases the ion inside the cell.

Active transport requires more energy than passive transport. Transport protein molecules assist substances to cross the membrane.

The most important active transport system in cells is that which moves sodium and potassium ions across cell membranes. The system is referred to as the **sodium–potassium pump**. Active transport keeps the inside of the cell high in potassium ion and low in sodium ion. Since the potassium ion concentration is lower outside the cell than in the cell cytoplasm, the transport of potassium ions into the cell takes place *against* the concentration gradient. In the case of sodium ion, the concentration is higher outside the cell than in the cell cytoplasm; therefore the movement of sodium ions out of the cell takes place *against* the concentration gradient. These ion concentrations are important in regulating the water content of the cell, in protein synthesis within the cell and in the operation of nerve cells and muscle cells. The protein in the phospholipid bilayer which takes part in the active transport of sodium ions and potassium ions is called Na^+–K^+–ATP-ase. [For ATP, see §9.5]. A large fraction of the ATP produced in the mammalian body at rest is used to maintain the distribution of sodium ion and potassium ion in the tissues.

The sodium–potassium pump is an important active transport system. It maintains the difference in the concentrations of sodium and potassium ions inside and outside the cells.

CHECKPOINT 4.12

1. The diagram represents the structure of a cell surface membrane.

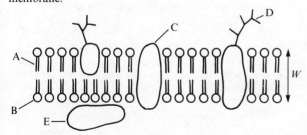

(a) State the names of components A–E.

(b) Give an approximate figure for the width W.

(c) Substances pass across a membrane by active transport and passive transport.

What is the difference between the two?

2. • Bacterial cell membranes are formed by phospholipids.
 • Soap has an anti-bacterial action.

Are these facts connected? Explain your answer.

3. (a) Describe the structure of a cell surface membrane.

(b) Explain how the structure of phospholipids allows them to form a bilayer.

(c) Explain why a cell surface membrane is (i) flexible and (ii) strong.

(d) When tissues form, cells of the same type associate. Explain how a cell of a certain type is able to recognise and bond to cells of the same type.

(e) Explain how a protein molecule which spans the cell surface membrane can act as a 'pore' through which substances may pass.

4.13 LECITHINS AND CEPHALINS

Lecithins and cephalins are complex lipids. Lecithins are widespread in animal and plant tissue and often make up the major portion of cell lipids. They occur in the brain, liver, egg yolk, soya bean, wheat germ and yeast. Cephalins are present in the nerve tissue of animals and in cell surface membranes. Lecithins have the general formula:

$$CH_2O_2CR^1$$
$$R^2CO_2CH \quad O$$
$$CH_2OPOCH_2CH_2N^+(CH_3)_3$$
$$O^-$$

Lecithin: $R^1CO_2^-$ and $R^2CO_2^-$ are long-chain fatty acid groups, one saturated and one unsaturated, and $—CH_2CH_2N^+(CH_3)_3$ is a choline group

Lecithins and cephalins are lipids. They are acylglycerols in which one of the acyl groups is a phosphoric acid ester.

A hydroxyl group of glycerol is esterified with a phosphate group which is in turn esterified with the base choline. The other two hydroxyl groups of glycerol form ester linkages with long-chain fatty acids, one saturated and the other unsaturated. (If, in the formula shown, R^1 is unsaturated and R^2 is saturated, this is called α-lecithin because the unsaturated alkyl group is in the 1-position or α-position.) Comparison with phosphoglycerides [§ 4.11] shows that lecithins have a choline group in place of the alkyl group of phosphoglycerides.

Cephalins are similar to lecithins, but contain the group $—CH_2CH_2NH_3^+$ in place of the choline group, $—CH_2CH_2N^+(CH_3)_3$.

4.14 EMULSIONS

Oil and water can be made to form a colloidal emulsion through the action of an emulsifier. An emulsifier has a hydrophilic polar group and a lipophilic non-polar group. The polar group dissolves in water, and the non-polar group dissolves in the oil.

'Oil and water don't mix', the saying goes. In milk, cream, mayonnaise and many sauces, however, you see mixtures of oil and water which are not separating into their components. These mixtures are **colloidal emulsions** [see *ALC*, § 9.5.5]. The oil is dispersed through the water as a suspension of tiny drops. The presence of an **emulsifier** makes this possible. An emulsifier combines in its structure a polar group which is hydrophilic (attracted to water) and lipophobic (repelled by lipids) and another group which is non-polar, hydrophobic and lipophilic [see Figure 4.14A]. Lecithin [§ 14.13 and Figure 4.14B] is widely used as an emulsifier. It is present in egg yolk; this is why egg yolk is a major ingredient in making mayonnaise and other salad dressings.

When the emulsifier is added to a mixture of oil and water, the emulsifier ions orient themselves so that the hydrophilic groups dissolve in water and the lipophilic groups dissolve in the oil. As a result, emulsifier ions arrange themselves round each droplet of oil [Figure 4.14C]. As the surface of each droplet is negatively charged, the drops repel one another, spread through the water, and do not coalesce.

FIGURE 4.14A
An Emulsifier Ion

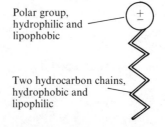

Polar group, hydrophilic and lipophobic

Two hydrocarbon chains, hydrophobic and lipophilic

FIGURE 4.14B
Lecithin

$N^+(CH_3)_3$
|
$(CH_2)_2$
|
O
|
$O=P-O^-$
|
CH_2 ⎫ Polar group, hydrophilic
|
$CHOCOCH_2CH_2CH_2R^2$ ⎫ Non-polar groups,
| lipophilic
$CH_2OCOCH_2CH_2CH_2R^1$ ⎭

FIGURE 4.14C
A Drop of Oil Surrounded
by Emulsifier Ions

For emulsifiers as food additives, see § 13.4. For the emulsifying action of soaps and soapless detergents on lipids, see *ALC,* § 33.13.3.

QUESTIONS ON CHAPTER 4

1. Give either a definition or an example of the following.

(a) lipid

(b) triacylglycerol

(c) unsaturated fatty acid

(d) phospholipid

(e) wax

(f) autoxidation

(g) antioxidant

2. Why does fish smell when it is not fresh?

3. Explain the difference between the members of each of the following pairs

(a) hydrolysis and hydrogenation

(b) hydrolysis and saponification

(c) phosphoglyceride and mixed triglyceride

(d) fat and oil

(e) wax and triacylglycerol

4. $CH_2OCO(CH_2)_{14}CH_3$
|
$CHOCO(CH_2)_{16}CH_3$
|
$CH_2OCO(CH_2)_7CH=CH(CH_2)_7CH_3$

(a) Name this triacylglycerol, which is found in lard.

(b) Write an equation for the formation of soap from this compound and potassium hydroxide.

5. Samples of soft margarine and hard margarine were heated. The percentage of solid fat in each sample was recorded, and the results tabulated.

Temperature/ °C	Solid fat/percent by volume	
	Soft margarine	Hard margarine
45	0	15
40	0	39
35	1	53
30	3	55
25	7	59
15	17	66
5	28	70

By means of graphs, find the melting temperature of each margarine. What does the difference tell you about the structures of the fats?

6. Refer to the formulae of the hormones oestrone and testosterone [§ 4.10].

(a) What is the major structural resemblance between them?

(b) What is the major structural difference between them?

(c) What substance is essential for mammals to use in the synthesis of many hormones of this type?

(d) What is the function of the bile acids?

5

NUCLEIC ACIDS

5.1 FUNCTION

The genetic material that determines the characteristics of an organism is DNA, deoxyribonucleic acid.

What is it that makes a man different from a mouse and a tiger different from a tortoise? The answer is not in the carbohydrates, lipids and proteins which make up the organisms; the answer lies in the **DNA** which the organisms contain. The DNA in an organism is the genetic material which decides what characteristics that organism will have and, with the help of **RNA**, what biochemical reactions will take place in it. Genetic information is transmitted from one generation to the next.

DNA and RNA were first identified in cell nuclei and are called **nucleic acids**. DNA is **deoxyribonucleic acid**, and RNA is **ribonucleic acid**. DNA and RNA are polymers, which have monomeric units called **nucleotides**. The monomeric units of DNA are **deoxyribonucleotides**, and the monomeric units of RNA are **ribonucleotides**. Each nucleotide contains three parts: a base, which is related to pyrimidine or purine, a pentose sugar and a phosphate group.

The bases in RNA and DNA are five compounds: three pyrimidine bases – cytosine, thymine and uracil – and two purine bases – adenine and guanine.

| Cytosine (in DNA and RNA) | Thymine (in DNA) | Uracil (in RNA) | Adenine (in RNA and DNA) | Guanine (in RNA and DNA) |

Pyrimidine bases   Purine bases

The sugars present in the nucleic acids are: D-ribose in ribonucleotides and 2-deoxy-D-ribose in deoxyribonucleotides [see § 3.2 for formulae]. In **nucleotides**, the base is bonded to carbon 1 of the ribose sugar and a phosphate group esterifies the hydroxyl group on carbon 5. A nucleoside is a compound of a purine or pyrimidine base with ribose; for example adenine and ribose combine to form adenosine:

64

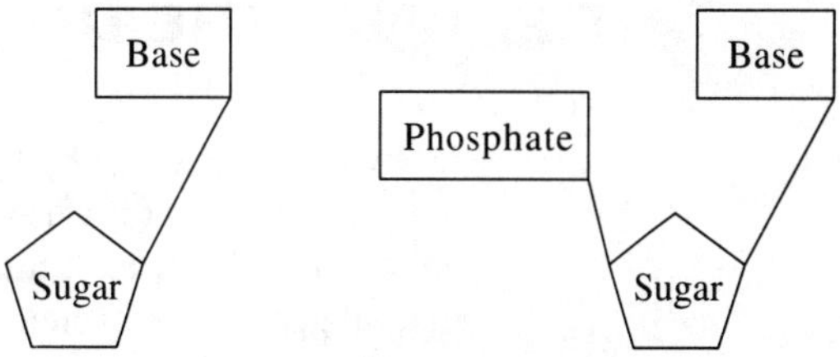

A nucleoside   A nucleotide

DNA and RNA, ribonucleic acid, are nucleic acids. DNA is a polymer of deoxyribonucleotides, and RNA is a polymer of ribonucleotides. A nucleotide molecule is composed of a base, a pentose sugar and a phosphate group. The bases are the pyrimidines cytosine, thymine and uracil and the purines adenine and guanine.

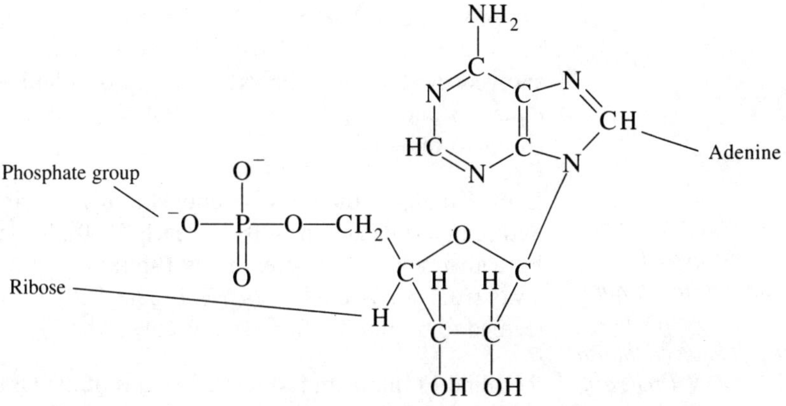

Adenosine monophosphate (a nucleotide)

Nucleotides polymerise to form nucleic acids [see Figure 5.1A].

FIGURE 5.1A
The Structure of a Nucleic
Acid

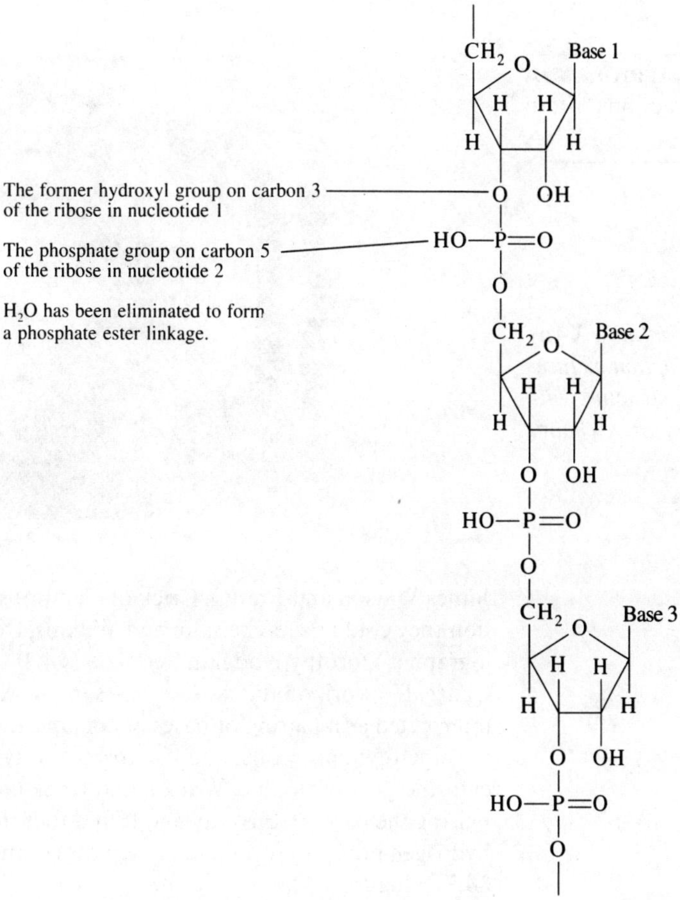

The former hydroxyl group on carbon 3 of the ribose in nucleotide 1

The phosphate group on carbon 5 of the ribose in nucleotide 2

H₂O has been eliminated to form a phosphate ester linkage.

Nucleotides polymerise by the formation of phosphate ester groups between a phosphate group on one nucleotide and a ribose hydroxyl group on another nucleotide.

5.2 DNA, THE DOUBLE HELIX

DNA was discovered by Friedrich Miescher in 1869 but aroused little interest. It was not until 1944 that it was suggested by O.T. Avery that DNA might be a very important substance: it might be the genetic material. Previously it had been thought that all genes were proteins. The interest which Avery's theory generated in the structure of DNA was intense.

The composition of DNA was known but not the three-dimensional shape of the molecule. DNA molecules from different cells have different values of relative molecular mass from 1×10^6 to 1×10^9. However, Erwin Chargaff showed in 1951 that the ratio (moles of adenine/moles of thymine) is always equal to one, and the ratio (moles of guanine/moles of cytosine) is always equal to one.

The structure of DNA has been elucidated as the culmination of many investigations. Chargaff showed that in DNA the ratio (Amount of adenine/ Amount of thymine) = 1 (Amount of guanine/ Amount of cytosine) = 1

Linus Pauling of the USA is famous for his contribution to our knowledge of the nature of chemical bonds. By the early 1950s, he had worked out that fibrous proteins have an α-helical structure; that is a spiral which turns in a clockwise direction going away from you [see *ALC*, §4.7.3, Figure 4.39]. He was investigating DNA, which seemed to be another fibrous molecule.

Rosalind Franklin and Maurice Wilkins of the University of London were tackling the problem by using X-ray crystallography. Franklin's X-ray diffraction pictures of 1952 were consistent with a helical structure with a diameter of 2 nm and a regular spacing of 0.34 nm along its axis. The patterns were suggestive of a structure involving two or three chains [see Figure 5.2A].

FIGURE 5.2A
X-ray Diffraction Pattern of DNA

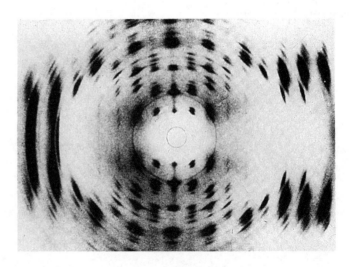

Franklin's X-ray diffraction pictures indicated a structure with two or three chains.

James Watson and Francis Crick of Cambridge University gleaned all the information they could from Franklin and Wilkins, from Pauling's son, from the crystallographer Dorothy Hodgkin [see *ALC*, §4.3] and others. Watson and Crick knew of Chargaff's work on the ratio of bases in DNA. They realised that it could be simply interpreted as a pairing of bases: if adenine always pairs with guanine, the ratio adenine/thymine = one, and if cytosine always pairs with guanine, the ratio cytosine/guanine = one. Watson and Crick constructed scale models. They tried pairing the bases in this way and found that the bases could be held together by hydrogen bonds to form pairs of equal size and shape [see Figure 5.2B and see *ALC*, §4.7.3, Figure 4.42].

FIGURE 5.2B
Hydrogen Bonding in
Adenine–Thymine and
Cytosine–Guanine Base
Pairs

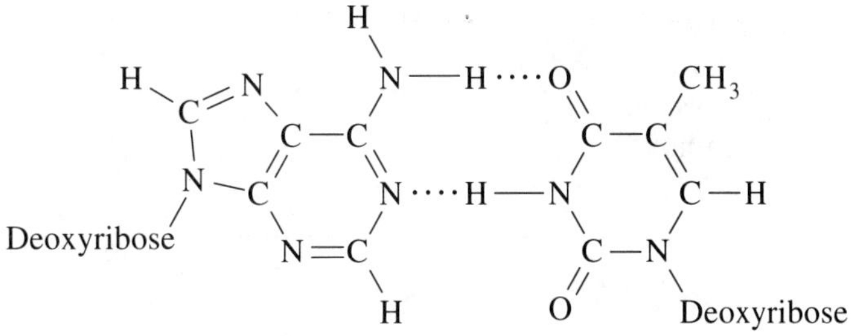

Adenine　　　　　　Thymine

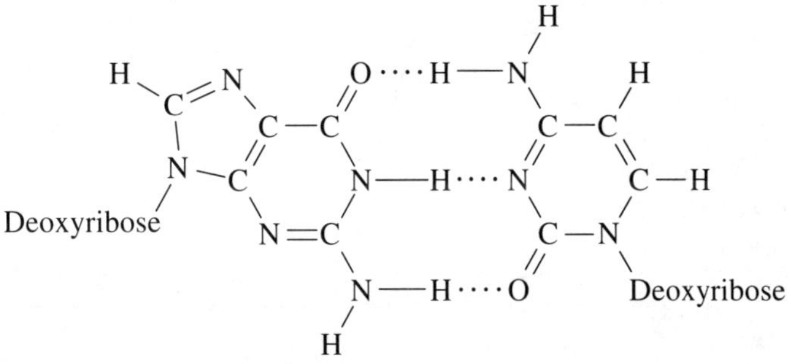

*Watson and Crick
suggested base pairing
through hydrogen bonding,
that is, adenine paired with
thymine and guanine
paired with cytosine.*

Guanine　　　　　　Cytosine

Watson and Crick constructed a model of a DNA molecule consisting of two separate polynucleotide chains [see Figure 5.2C]. In the three-dimensional structure, each chain has the form of a right-handed helical spiral, and the two chains coil round each other to form a double helix [see Figure 5.2D]. Each chain has a sugar–phosphate backbone from which the bases project at right angles. Hydrogen bonds form between an adenine base in one strand and a thymine base in the other strand, and between guanine in one strand and cytosine in the other strand. If the sequences of bases in a length of one strand is AGTC, the sequence in the same length of the other strand is TCAG. The two strands are said to be **complementary**. The distance between the two chains is equal to the width of a purine base plus a pyrimidine base. The distance between base pairs in the scale model was 0.34 nm, which fitted in perfectly with Franklin's X-ray diffraction pictures [see Figure 5.2A]. The helix does a complete turn in ten base pairs, 3.4 nm.

At pH 7, the phosphate groups are ionised and negatively charged. Repulsion between adjacent phosphate groups tends to open out and lengthen the spiral. This repulsion is counteracted by magnesium ions which help to keep DNA in helical form. Watson and Crick published their structure in 1953. Their model was a triumph because it explains how DNA can act as the genetic material. It explains how DNA carries information for controlling protein synthesis [see § 5.4] and how DNA is able to replicate itself (make exact copies of itself) in order to pass the genetic information to the next generation [see § 5.5].

*They constructed a model
in which two α-helical
chains of polynucleotides
coiled round one another.
Base pairing held the
strands together. This
structure is the famous
'double helix'.*

The story of this discovery is recounted by James Watson in his book, *The Double Helix*. Together with Wilkins, Watson and Crick were awarded the Nobel prize in 1962. Nobel prizes go only to the living. Rosalind Franklin, on whose X-ray pictures the structure was based, had died and received no prize.

FIGURE 5.2C
Structure of DNA
Showing Straightened
Chains

*Two polynucleotide chains
are joined by hydrogen
bonds between adenine and
thymine and between
guanine and cytosine.*

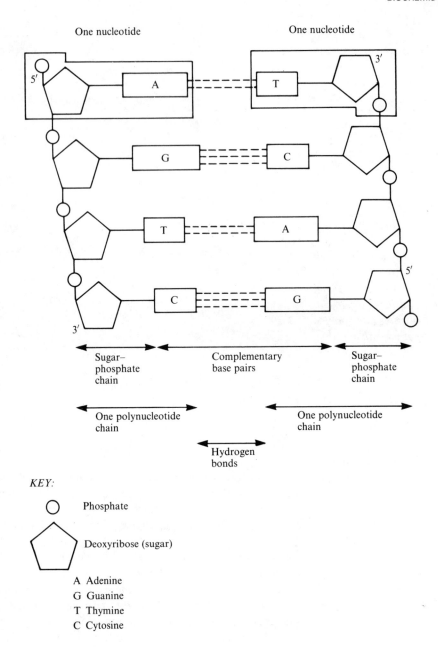

FIGURE 5.2D
The Helical Structure of
DNA in Outline

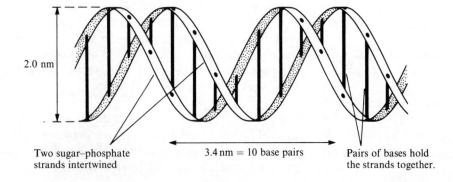

FIGURE 5.2E
Part of the DNA Double
Helix

FIGURE 5.2E
Part of the DNA Double
Helix

*The two polynucleotide
chains coil round one
another as two parallel
linked spirals – the double
helix.*

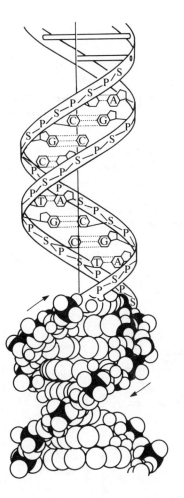

FIGURE 5.2F
James Watson and
Francis Crick with their
Model of DNA

Base pairs   C   P   O   H

CHECKPOINT 5.2

1. Copy and complete the following account.
Nucleotides are organic compounds containing the
elements _____ . A molecule of the mononucleotide
ATP includes the base _____ and the
sugar _____ , which has _____ carbon atoms.
The organic base has a double-ringed molecule called
a _____ . RNA and DNA both contain a nucleo-
tidewhich includes the base guanine, but in RNA the
nucleotide contains one more atom of _____ . In DNA,
the nucleotide containing guanine pairs with the nucleotide
containing _____ by the formation of _____
bonds.

2. (*a*) Name (i) the purines, (ii) the pyrimidines in DNA.

(*b*) Why is a molecule of DNA described as a 'double helix'?

(*c*) What is meant by the statement that the two strands of
DNA are complementary?

(*d*) Explain how the polynucleotide chains are held together
in the double helix of DNA.

5.3 RNA

RNA consists mainly of single-stranded molecules. The bases include uracil in place of thymine. The sugar is ribose.

There are three types of RNA. All of them have a part to play in protein synthesis which is described in the next section. Molecules of RNA usually consist of single strands, but transfer RNA has some regions where base-pairing occurs. The bases in RNA are adenine (A), uracil (U), guanine (G) and cytosine (C), and the sugar is ribose. The three types of RNA are as follows:

1. **Messenger RNA (mRNA)** contains the bases A,G,C and U. mRNA is synthesised in the nucleus. Its molecules vary from $M_r = 25\,000$ to $1\,000\,000$. The molecules of mRNA are complementary to a portion of a DNA molecule.

2. **Transfer RNA (tRNA)** is of different types, of $M_r = 23\,000–30\,000$. The function of tRNAs is to act as 'carriers' of amino acids during protein synthesis. There is one tRNA which is able to bond to each one of the 20 amino acids in proteins. tRNA molecules have the structure shown in Figure 5.3A. They are held in this configuration by base pairing within the molecule.

FIGURE 5.3A
The Structure of Transfer RNA, tRNA. Note the 'clover leaf' structure

RNA includes messenger RNA (mRNA) transfer RNA (tRNA), of which there are many different kinds, each adapted to transfer a different amino acid, . . .

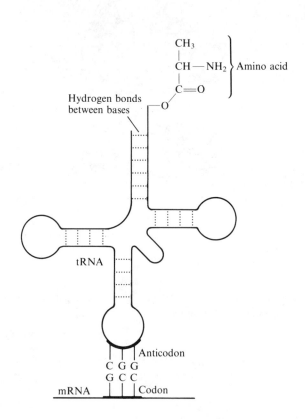

. . . and ribosomal RNA, rRNA.

3. **Ribosomal RNA (rRNA)** is a structural component of ribosomes. The molecules consist of single strands and have M_r values up to $1\,000\,000$.

All three types of RNA are synthesised on DNA. In order to form mRNA, a double helix of DNA unwinds. One of the single strands of DNA acts as a template (pattern). The bases in this strand of DNA attract ribonucleotides which contain complementary bases. The ribonucleotides which have bonded to the template strand of DNA combine with each other to form a strand of mRNA [see Figure 5.3B]. The process is called **transcription** because the information in DNA has been transcribed (copied) on to the new mRNA.

FIGURE 5.3B
Transcription of RNA
from DNA

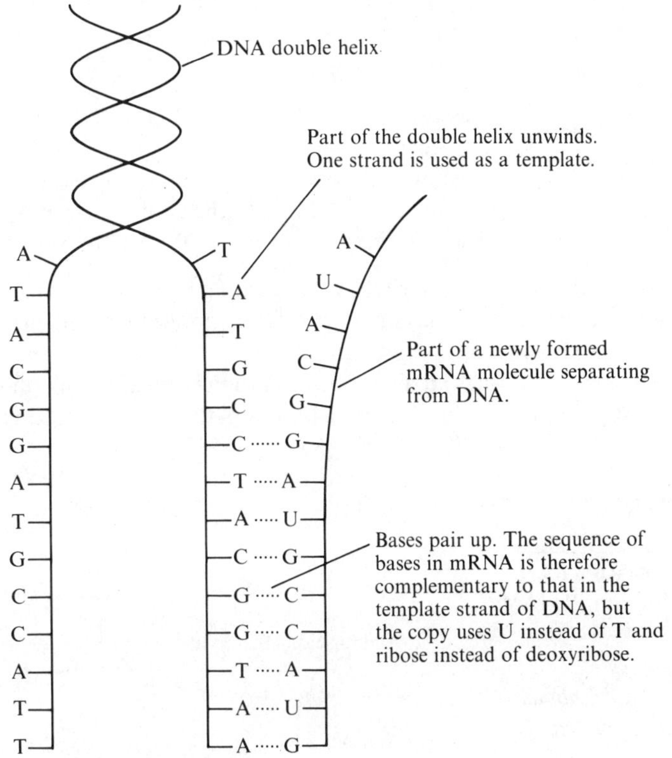

DNA double helix

Part of the double helix unwinds.
One strand is used as a template.

Part of a newly formed
mRNA molecule separating
from DNA.

Bases pair up. The sequence of
bases in mRNA is therefore
complementary to that in the
template strand of DNA, but
the copy uses U instead of T and
ribose instead of deoxyribose.

*mRNA is synthesised on
DNA by transcription.
One strand of DNA acts as
a template to which free
ribonucleotides bond by
base pairing. The
information on DNA has
been transcribed to
mRNA.*

5.4 PROTEIN SYNTHESIS

The synthesis of protein is one of the activities which take place within cells. To make
one molecule of a certain protein, hundreds or thousands of amino acid molecules
must combine in the correct sequence. This feat is achieved through the action of
DNA and RNA. The sequence of bases, A, T, G, C, in DNA carries instructions,
called a **code**, which make amino acids combine in the correct order. The instructions
needed to place one amino acid in its correct position in the protein molecule are
contained in a row of three bases, called a **triplet code**. Why is there a triplet code for
each amino acid? You can see that four individual bases could only code for four
amino acids. The number of base pairs that can be formed from four bases is $4^2 = 16$,
which is not enough to code for 20 amino acids. The number of ways of combining
four bases in triplets is $4^3 = 64$, which is more than enough to code for 20 amino
acids. No two amino acids have the same triplet code. One amino acid may be coded
for by more than one triplet code. The code is universal for all forms of life from
bacteria to humans.

*The synthesis of a protein
requires hundreds of amino
acids to combine in the
right order. The sequence
of bases on DNA carries
the information needed to
assemble amino acids in the
correct order. A triplet
code of three bases codes
for each amino acid.*

*A length of DNA called a
gene codes for a whole
protein.*

A length of DNA which codes for a whole protein is called a **gene**, and the DNA code
is called the **genetic code**. Proteins are synthesised on the ribosomes in the cytoplasm.
Since DNA is present in the nucleus, there must be a mechanism which enables it to
control events in the cytoplasm. Instructions must be able to travel from DNA in the
nucleus to the ribosomes in the cytoplasm. The instructions are carried by **messenger
RNA (mRNA)**. Messenger RNA can travel out of the nucleus through pores in the
nuclear membrane into the cytoplasm, where it bonds to a **ribosome**. Because of the
way it is made [see Figure 5.3B], mRNA consists of a sequence of nucleotides which
are the same as the triplet codes on one strand of DNA. A triplet of nucleotides in
mRNA is called a **codon**. The codons ACU, ACC, ACA and ACG code for

*The genetic information on
DNA is transcribed to
mRNA.*

mRNA passes out of the nucleus and bonds to a ribosome. The triplets of bases on mRNA are called codons.

threonine. The codons GAU and GAC code for aspartic acid. The codon AUG will initiate a peptide chain, and codons UAA, UAG and UGA will terminate a polypeptide chain.

The function of ribosomes is to **translate** the sequence of codons on mRNA into a sequence of amino acids in a polypeptide chain. There is a difficulty to be overcome because amino acids cannot bond to codons. An intermediary is needed to bind to a codon at one end and to the aminoacid for which the codon codes at the other end. The intermediary is **transfer RNA (tRNA)** [Figure 5.4A].

The ribosomes translate the sequence of codons on mRNA into a sequence of amino acids in a polypeptide chain.

Transfer RNA has a vital part to play. It is able to bind to amino acids and to RNA. At one end of each tRNA molecule is a triplet of unpaired bases, an **anticodon**, that is the complement of a particular RNA codon. At the other end of the molecule, tRNA has an amino-acid-binding site which will bind the amino acid for which the codon codes. Figure 5.4A shows the steps in protein synthesis.

FIGURE 5.4A
Translation of mRNA

Each kind of tRNA bonds to a certain amino acid.

1 tRNA bonds to its specific amino acid.

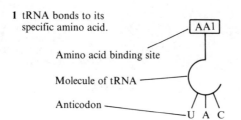

Amino acid binding site

Molecule of tRNA

Anticodon

2 mRNA has travelled out of the nucleus into the cytoplasm and bonded to a ribosome.

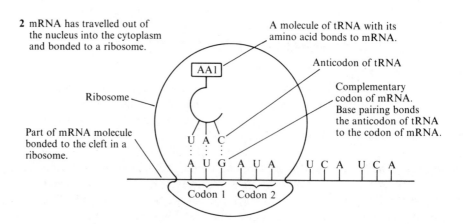

A molecule of tRNA with its amino acid bonds to mRNA.

Anticodon of tRNA

Complementary codon of mRNA. Base pairing bonds the anticodon of tRNA to the codon of mRNA.

Ribosome

Part of mRNA molecule bonded to the cleft in a ribosome.

3 There are three binding sites on the ribosome, a cleft for mRNA and two sites for tRNA. A second molecule of tRNA can therefore bind to the second codon on the mRNA. Amino acid 2 combines with amino acid 1 to form a dipeptide.

The ribosome moves along a strand of mRNA. As it reaches each codon on the mRNA, a molecule of tRNA, bonded to the amino acid for which that codon codes, transfers its amino acid to the ribosome.

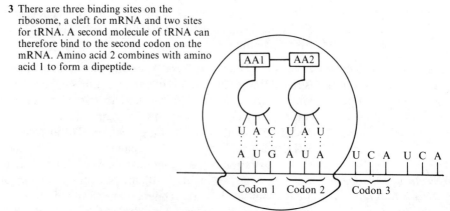

FIGURE 5.4A
Translation of mRNA
continued

4 The ribosome moves to a
third codon on mRNA.
A third tRNA combined
with its amino acid bonds
to it by base pairing.
Amino acid 3 combines
with amino acid 2.
The first tRNA molecule
is released.

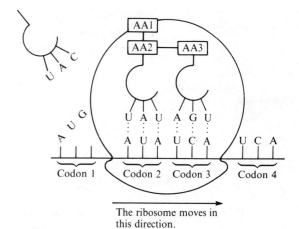

The ribosome moves in
this direction.

5 As the process continues, a polypeptide
chain is built up. The order in which
amino acids combine is determined by the
code in mRNA, which is a complementary
copy of the code in one of the strands of DNA.
The process is called **translation** of mRNA.
The polypeptide chain is finally released
when mRNA separates from the ribosome.

*The amino acids brought to
the ribosome join to form a
polypeptide. When
complete, the polypeptide
is released from the
ribosome.*

The polypeptide chain folds in a characteristic manner to acquire the three-
dimensional structure of the protein. It is released when mRNA finally becomes
detached from the ribosome. By controlling protein synthesis, including the synthesis
of enzymes, DNA controls all the functions of the cell.

5.5 REPLICATION OF DNA

Cells can divide to form two new cells. The cells which divide are called **parent cells**,
and the new cells are called **daughter cells**. In cell division, the nucleus divides and the
cytoplasm divides. There are two ways in which the nucleus can divide: mitosis and
meiosis. In **mitosis**, the two daughter cells have the same number of chromosomes as
the parent cell and are therefore genetically the same as the parent cell and one
another. Mitosis enables living things to grow, as when plants lengthen their roots,
and to repair damaged tissues, as when new skin grows over a wound. In **meiosis**, each
daughter cell has half the number of chromosomes which the parent cell possessed.
This method of cell division is employed in the production of gametes (sex cells).

Before mitosis takes place, new DNA is made in the nucleus. Each double helix of DNA makes an exact copy – a replica – of itself, and the process is called **replication** [see Figure 5.5A]. The double helix structure of DNA enables this to happen.

FIGURE 5.5A
Replication of DNA:
Semiconservative
Replication

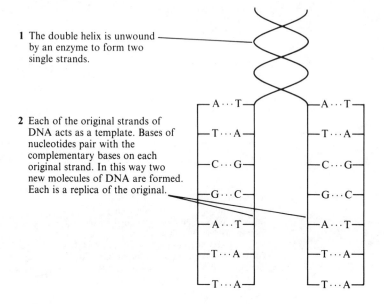

1 The double helix is unwound by an enzyme to form two single strands.

2 Each of the original strands of DNA acts as a template. Bases of nucleotides pair with the complementary bases on each original strand. In this way two new molecules of DNA are formed. Each is a replica of the original.

5.5.1   SEMICONSERVATIVE REPLICATION

Two suggestions concerning the role of DNA as template have been put forward. Watson and Crick suggested that each chain of the DNA double helix serves as a template for the synthesis of a complementary chain. This model is known as **semiconservative replication**. It predicts that the template DNA molecule will be divided equally between the two newly synthesised molecules after one replication [see Figure 5.5A]. The other suggestion is **conservative replication**. It is visualised that the DNA double helix serves as template for the duplication of only one of the DNA strands. The newly formed strand subsequently serves as template for the formation of its complementary strand. In this case, after one replication the original DNA molecule would remain intact and the new DNA molecule would consist of two new strands [Figure 5.5B].

When cells divide by mitosis, the daughter cells have the same number of chromosomes as the parent cell. Before cell division, the cell DNA replicates itself.

The set of experiments carried out by Meselson and Stahl distinguished between the two suggested mechanisms of replication. These workers grew *E. coli* cultures on a medium containing isotopically pure $^{15}NH_4Cl$ as the only source of nitrogen. The DNA synthesised in the course of cell multiplication is therefore 'heavy': it contains ^{15}N, rather than the normal ^{14}N. 'Heavy' DNA can be distinguished from 'light' DNA by the rate at which it sediments in a high-speed centrifuge called an **ultracentrifuge**.

Is the mechanism conservative replication or semiconservative replication?

Cultures grown on $^{15}NH_4Cl$ were transferred to a medium containing the normal $^{14}NH_4Cl$, and the DNA of these cells was examined in the ultracentrifuge at intervals over a period of time [see Figure 5.5C]. At first, only one band of sediment, 'heavy' DNA, was observed. After one generation of growth in the ^{14}N medium, the DNA from *E. coli* showed as a single band of sediment. The density was midway between those for 'heavy' DNA and 'light' DNA. This showed that after a single replication the daughter molecules must have contained one strand of the original 'heavy' DNA and one strand of newly synthesised 'light' DNA. This result fits in with the semiconservative model of replication.

The Meselson–Stahl experiments provide evidence.

Chromosomes consist of strands of DNA coiled round a protein core [Figure 5.5D].

FIGURE 5.5B
Conservative Replication of DNA

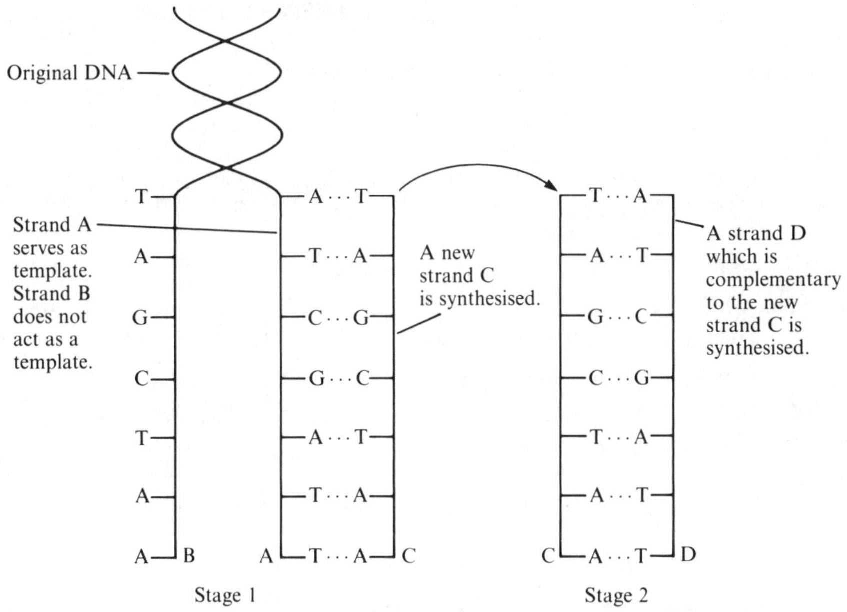

Original DNA
Strand A serves as template. Strand B does not act as a template.
A new strand C is synthesised.
A strand D which is complementary to the new strand C is synthesised.
T
A
G
C
T
A
A
B
A···T
T···A
C···G
G···C
A···T
T···A
T···A
C
T···A
A···T
G···C
C···G
T···A
A···T
A···T
D
Stage 1
Stage 2

The mechanism is semiconservative replication: each strand of the double helix serves as a template for the synthesis of a complementary chain.

FIGURE 5.5C
DNA Sedimentation Patterns

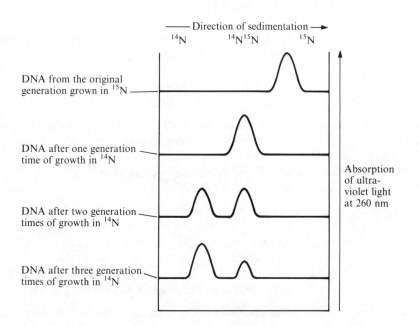

Direction of sedimentation
^{14}N
$^{14}N^{15}N$
^{15}N
DNA from the original generation grown in ^{15}N
DNA after one generation time of growth in ^{14}N
DNA after two generation times of growth in ^{14}N
DNA after three generation times of growth in ^{14}N
Absorption of ultra-violet light at 260 nm

The alternative mechanism, conservative replication, has been ruled out by experiments with ^{15}N-labelled DNA.

FIGURE 5.5D
The Structure of a Chromosome

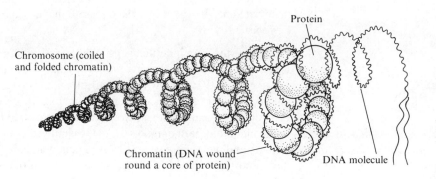

Protein
Chromosome (coiled and folded chromatin)
Chromatin (DNA wound round a core of protein)
DNA molecule

1. What is the difference between a codon and an anticodon?

2. What is the difference between transcription and translation?

3. (*a*) Describe the characteristics of DNA that allow self-replication to take place.

(*b*) What chemical evidence indicates that the replication of DNA is semi-conservative?

4. (*a*) Name the four bases in DNA.

(*b*) Explain how base pairs are bonded together.

(*c*) What is transcription? How does it happen?

(*d*) What is the function of tRNA?

5. (*a*) Name the three components that make up a nucleotide.

(*b*) What is the difference between a nucleotide and a nucleic acid?

(*c*) List three functions of nucleotides in the cell.

6. Meselson and Stahl performed an experiment on *E. coli.* They grew cells for several generations in the presence of ^{15}N. This procedure labelled the DNA of progeny that were produced from the initial small sample with ^{15}N. They placed the ^{15}N-labelled cells in a ^{14}N-containing medium. After one generation time (50 min at 36 °C for *E. coli*), they extracted the DNA. They repeated the extraction after other intervals of one generation time. The DNA from different generations was compared by finding the rate of settling in an ultra-centrifuge. The positions of ^{15}N-DNA, ^{14}N-DNA and ^{14}N-^{15}N-DNA (DNA containing both isotopes) were found from the absorbance of ultraviolet light at 260 nm [see Figure 5.5C].

Do the results shown in the figure indicate that the replication of the DNA is conservative or semiconservative? Give reasons for your answers

5.6 MUTATIONS

When DNA replicates, occasionally a wrong nucleotide adds by mistake. The change is called a genetic mutation. Genetic mutations give rise to a number of inherited diseases:

When DNA replicates, hundreds of nucleotides add each second to the growing strand of DNA. Sometimes the wrong nucleotide adds by mistake, and the new DNA is slightly different from the original. This change is called **mutation**. The chance of a genetic mutation occurring is increased by irradiation from X-rays and radioactive sources and by some chemicals [see *ALC*, § 30.13].

Sickle cell anaemia is a disease which occurs in Africa and in the black population of the USA. The red blood cells are abnormally shaped, many having elongated shapes or sickle-shapes or crescent-shapes. The mis-shapen cells become trapped in blood vessels, and impair the circulation of the blood, resulting in diseases of the liver and kidney and heart attacks.

...sickle cell anaemia, in which blood circulation is impaired...

The structure of haemoglobin is shown in Figure 2.9A. Research has shown that in sickle cells a β-subunit of haemoglobin has a valine amino acid residue in place of the normal glutamic acid residue. While glutamic acid is highly polar, valine is non-polar, and the change makes a big difference to the solubility of sickle cell haemoglobin. The sparingly soluble haemoglobin molecules clump together to form fibrous masses which distort the shape of the red blood cells.

...phenylketonuria, which can result in mental retardation...

Phenylketonuria is an inherited disease caused by a mutation in which the enzyme which converts the amino acid phenylalanine into tyrosine is absent. As a result phenylalanine accumulates in the body fluids. The result is mental retardation and a short life expectancy. The disease can be treated by supplying a diet that is low in phenylalanine, provided that the disease is diagnosed early by a blood test on a newborn baby.

...galactosaemia, in which milk cannot be digested.

Galactosaemia is an inherited genetic disease. Children who suffer from the disease lack an enzyme which converts the sugar galactose in milk into glucose. The children fail to grow and are sick when they drink milk. The results are mental retardation and, if the child continues to be given galactose, death can follow. The disease can be controlled by means of a galactose-free diet, provided that the absence of the enzyme in the blood has been diagnosed. As the child grows older, other enzyme systems develop and metabolise galactose. A person who inherits galactosaemia can therefore have a normal adult life provided the condition is diagnosed and treated in childhood.

━━━━━━━━━━━━━━━━━ **CHECKPOINT 5.6** ━━━━━━━━━━━━━━━━━

1. (*a*) Why is it important that DNA replication should produce two exact copies of the original DNA molecule?

(*b*) What is the result of the occasional mistake in replication?

2. (*a*) Explain what is meant by an inherited genetic disease.

(*b*) Give two examples of such diseases.

(*c*) Describe one way in which genetic disease can arise.

5.7 ISOLATION OF NUCLEIC ACIDS FROM CELLS

Nucleic acids are obtained from cells by denaturing the protein content, partitioning the cell contents between aqueous phenol and water, and precipitating nucleic acids from the water layer.

Nucleic acids can be separated from biological materials. In the cell, nucleic acids are often bound to proteins, and techniques must be employed to separate them. A much-used method is to grind the tissue in a buffered solution of detergent. This denatures the protein. A saturated solution of phenol is added and the extract is stirred for a time and then centrifuged. Two layers form. The lower layer, the phenol layer, contains denatured proteins and other parts of the cell. The upper layer, the aqueous layer, contains nucleic acids in solution. When ice-cold ethanol is added to this layer, the nucleic acids are precipitated from solution.

The nucleic acids must be purified. Carbohydrates are removed by chromatography. DNA must be separated from RNA. If DNA is required, an enzyme is used to hydrolyse the RNA in the extract. If RNA is required, the DNA in the extract is hydrolysed by another enzyme. The different types of RNA are separated by means of their different solubilities in concentrated salt solutions.

DNA profiling is used to identify a person from their DNA.

DNA profiling, popularly known as **genetic fingerprinting** is a technique which can identify a person from a sample of their genetic material. Human DNA contains sequences of bases which are repeated many times. The number and pattern of repeats is unique in each individual. These sequences can be made visible by labelling the DNA with radioactive phosphorus-32 and exposing a photographic film to the sample. The DNA profiles so obtained are different for every person, except for identical twins. Genetic fingerprinting can be used on small samples of blood and semen to establish whether a suspect was at the scene of a crime. It can be used to identify whether a man and woman are the parents of a child. In Argentina, a number of abducted children have been identified and reunited with their families by means of genetic fingerprinting.

5.8 GENETIC ENGINEERING

Genes can be transferred from one organism to another. The technique is called recombinant DNA technology, or gene manipulation or gene cloning.

In the 1970s and 1980s, biochemists found out how to alter genes and how to transfer them from one organism to another. They used the technique to alter the DNA of rapidly reproducing organisms such as bacteria. They were able to change the DNA of micro-organisms in a specific way so as to change the proteins which they produce. They could then use the micro-organisms as 'biochemical factories' to produce valuable proteins. This process of altering the genes is called **genetic engineering**. The method is to insert a strand of DNA, e.g. human DNA, into the DNA of a micro-organism, usually *E. coli*. The micro-organism will grow, and when it reproduces it will automatically copy the 'alien' DNA along with the rest of its DNA and will manufacture the protein for which the 'alien' DNA codes. If the protein is an enzyme it will catalyse reactions which can result in the accumulation of other interesting products of metabolism. Genetically engineered bacteria have produced new medicines and foods and fuels.

The method is to implant a human gene in a bacterium which reproduces rapidly. The result is a strain of genetically engineered bacterium. The bacterium synthesises the protein for which the human gene codes.

The potential of this discovery excited many scientists. They could see the possibility that the designed or engineered organisms might have a new range of properties; for example, they might be able to degrade plastics or they might be able to manufacture valuable proteins such as the hormone insulin. Biochemists predicted that micro-organisms might be genetically improved to make them produce large quantities of protein for consumption [see § 15.3]. This was achieved with ICI's 'Pruteen' a protein which is synthesised by a genetically engineered micro-organism. Genetic engineering is also called **genetic manipulation** or **gene cloning** or **recombinant DNA technology**.

Genetic engineering employs three techniques, each of which depends on the use of specific enzymes.

1. Cutting DNA into sections

The first step is to employ restriction endonucleases to cut human DNA into sections. Some sections contain intact genes.

Enzymes are used to cut DNA into sections. These enzymes, called **restriction endonucleases**, occur in several species of bacteria. Different endonucleases cut through DNA at different base sequences. Some leave sections of DNA with a number of unpaired nucleotide bases at the ends, which are called 'sticky ends'. These can later be joined with other fragments of DNA with 'sticky ends' in which the base sequences are complementary. Some of the sections of DNA produced contain intact genes. The first step in genetic engineering is therefore to use a restriction endonuclease to cut a segment of DNA from a chromosome, e.g. a human chromosome [see Figure 5.8A].

2. Joining segments of human DNA and bacterial DNA

Next a segment of human DNA is inserted in a bacterial plasmid (a piece of bacterial DNA separate from the bacterial chromosome). The plasmid vector (plasmid + new DNA) is restored into the bacterial cell.

Bacterial cells contain small circular loops of DNA called **plasmids**. They are distinct from the rest of the DNA which makes up the bacterial chromosome, but they replicate in step with it. Plasmids are not essential for the survival of the organism. In this technique a plasmid is removed from a bacterial cell. The plasmid is cut open using the same restriction endonuclease as before so that the 'sticky ends' will be complementary with those of the DNA segment from Stage 1. The segment of alien DNA is inserted into the cut plasmid. The complementary bases in the 'sticky ends' pair up. An enzyme called a **DNA ligase** is used to seal the join. The alien DNA has now become part of the plasmid, which is now called a **plasmid vector**. The plasmid vector is restored into the bacterial cell, the **host cell**, thus acting as a **carrier** or **vector** for the gene [see Figure 5.8A]. In some cases, a virus is used as carrier instead of a plasmid.

3. Replication of DNA

Bacteria grow and divide rapidly. Each time the host organism divides, it replicates the foreign DNA along with the rest of its DNA. The bacterium is grown to produce a culture of identical cells (a clone) which all contain the new DNA.

The transfer of DNA to bacterial cells is successful in only a small proportion of cells. In order to select these cells, the plasmid carrier is given a **marker gene**. This is often a gene which confers resistance to a certain antibiotic, e.g. tetracycline. If the bacterium is cultured on a medium containing tetracycline, only the transformed cells will be able to grow. The successfully transformed cells are separated and grown on an industrial scale in fermenters. They manufacture the protein for which the new gene codes.

As a result of this process, a clone of bacteria which contain a portion of human DNA including a human gene has been produced. Other sections of DNA are introduced into bacteria and replicated in the same way. In this way a bank of bacterial clones, each containing a different human gene, is built up. Geneticists can study the different clones and select those that have the genes they want to use.

FIGURE 5.8A
Recombinant DNA
Technology

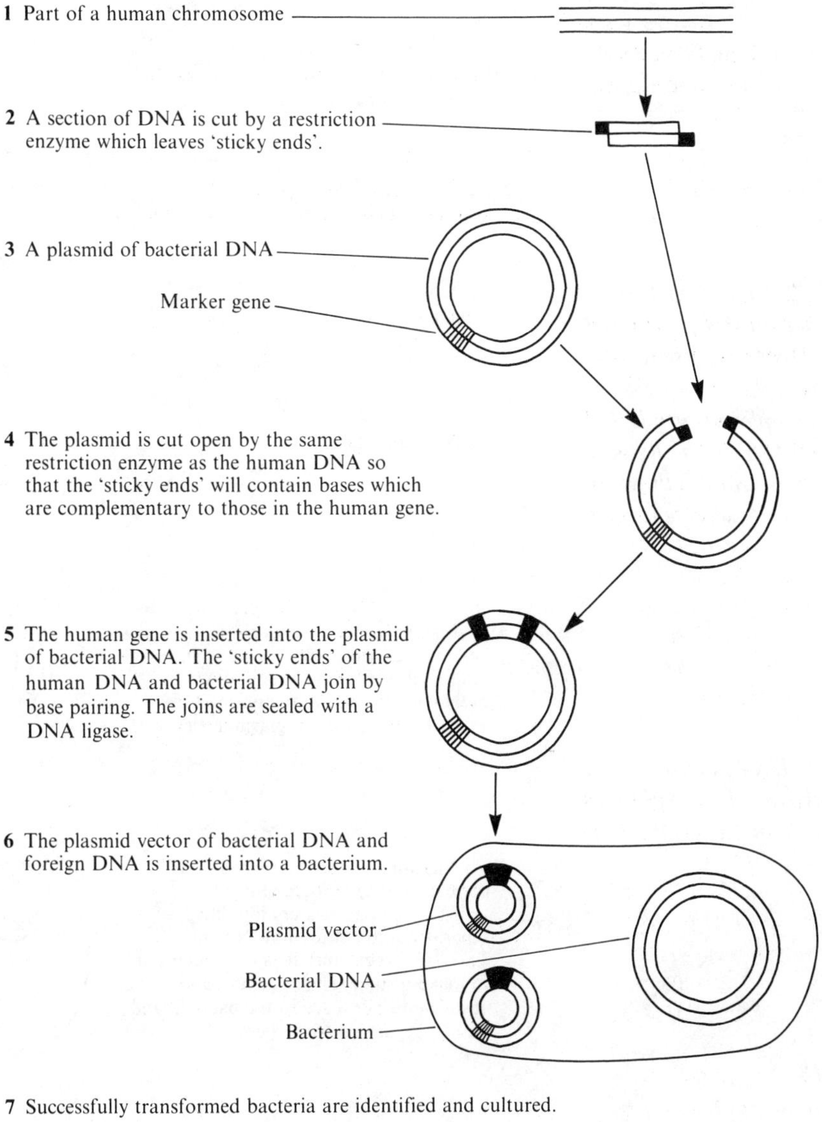

1 Part of a human chromosome ———————————

2 A section of DNA is cut by a restriction ——
enzyme which leaves 'sticky ends'.

3 A plasmid of bacterial DNA——

Marker gene——

4 The plasmid is cut open by the same
restriction enzyme as the human DNA so
that the 'sticky ends' will contain bases which
are complementary to those in the human gene.

5 The human gene is inserted into the plasmid
of bacterial DNA. The 'sticky ends' of the
human DNA and bacterial DNA join by
base pairing. The joins are sealed with a
DNA ligase.

*The bacterium divides and
replicates its DNA. A clone
of bacterial cells is formed,
each containing the new
human DNA.
A marker gene, previously
given to the plasmid,
enables successfully
transformed bacteria to be
selected.*

6 The plasmid vector of bacterial DNA and
foreign DNA is inserted into a bacterium.

Plasmid vector ——

Bacterial DNA ——

Bacterium ——

7 Successfully transformed bacteria are identified and cultured.

5.8.1 INSULIN

*An example of genetic
engineering is the use of
genetically engineered
bacteria to synthesise
insulin.*

A dramatic contribution of genetic engineering was the production of the human
hormone insulin by genetically engineered bacteria. In 1980 volunteer diabetics tried
out the genetically engineered insulin, and by 1982 it was in general use. Before then,
insulin was obtained from slaughtered cattle and pigs. It was expensive to produce and
some diabetics reacted allergically to animal insulin. Genetically engineered insulin is
cheaper, available in large quantities and chemically the same as human insulin. The
process was developed by Eli Lilly, and their product is known as Humulin.

There is more than one technique for duplicating portions of DNA. The β-cells in the
islets of Langerhans in the pancreas produce insulin, and these cells therefore contain
a relatively large amount of the messenger RNA that codes for insulin. The enzyme
reverse transcriptase can be used to synthesise DNA from the RNA in pancreas cells.
(This is the reverse of the transcription of RNA from DNA; see §5.3). The DNA
made in this way is called complementary DNA, cDNA (or copy DNA). A large
proportion of the cDNA produced is likely to code for insulin [see Figure 5.8B].

FIGURE 5.8B
Making Genetically
Engineered Insulin

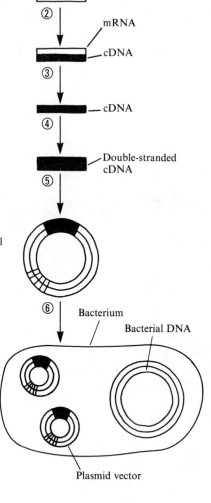

1 Messenger RNA, mRNA, is extracted
from human pancreas cells

2 Reverse transcriptase catalyses the
synthesis of complementary DNA, cDNA.

*A different method of
duplicating portions of
DNA is to use reverse
transcriptase to synthesise
complementary DNA,
cDNA, from the RNA in
the pancreatic cells which
secrete insulin.*

3 Alkali removes mRNA.

4 cDNA replicates.

5 A plasmid has been removed from a bacterial
cell and opened up by a restriction
endonuclease (as in Figure 5.8A)
cDNA is attached to the plasmid by
DNA ligase to form a plasmid vector.

*The cDNA replicates and
is introduced into a plasmid
of bacterial DNA.*

6 The plasmid vector is returned into
the bacterial cell. Successfully
transformed bacteria are identified,
separated and cultured in fermenters.
The insulin gene instructs the bacterium
to make insulin. The large amounts of
insulin are of no use to the bacteria and
can be separated from them.

*Among the products of
genetic engineering are
human growth hormone . . .*

In one child in 5000, too little growth hormone is secreted, and the child suffers from
dwarfism. Growth hormone from animals is ineffective in humans, and human
growth hormone can be obtained only from the pituitary glands of human corpses. As
a result, the hormone is in short supply. Human growth hormone has now been
produced by genetic engineering and is undergoing clinical trials.

*. . . and vaccines for
hepatitis B and for foot and
mouth disease in cattle.*

Other genetically engineered hormones include the hormone that controls the
absorption of calcium into bones. Vaccines for hepatitis B and foot-and-mouth
disease in cattle have been produced by genetic engineering. Intensive research is
underway in the hope of finding vaccines for malaria and AIDS.

*The techniques of genetic
engineering are costly and
problematic, and research
and development continue.*

The techniques of genetic engineering have proved more costly and more
problematic than foreseen. The host micro-organism may reject the inserted DNA,
excising it or inactivating it. The 'alien' proteins which are synthesised may be
modified by the cellular mechanisms of the micro-organism. Large-scale purification of proteins from the bacterial broth in which they are synthesised may not be
simple. The difficulties of applying recombinant DNA technology are very much
greater in multicellular organisms. Sexual reproduction makes the perpetuation of
artificially inserted DNA much more problematic. However, there have been some

successes, e.g. the insertion of the gene for human or rat growth hormone into mice has produced larger and heavier mice.

QUESTIONS ON CHAPTER 5

1. Write an essay on the elucidation of the structure of DNA.

2. (*a*) The base sequence of part of a DNA strand is: TACGGATTCATG. Write the base sequence of the mRNA which is transcribed from this sequence.

Explain the sequence of events that occurs during protein synthesis. The diagram will help you.

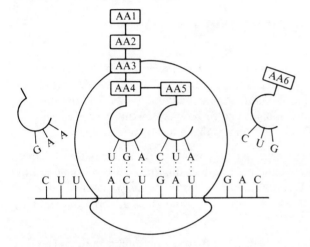

3. (*a*) Describe the structure of DNA.

(*b*) Explain how DNA acts as a code for protein manufacture.

4. Explain what is meant by *genetic engineering*. Give two examples of the use of this technique.

5. What are the three types of RNA found in the cell? Describe the function of each type.

6. 'The two strands of DNA are not identical; they are complementary'. Explain the meaning of this statement.

7. What are the three main structural differences between a molecule of DNA and a molecule of RNA?

8. Explain how genetic information is carried by the DNA molecule.

9. Give an illustrated account of the mechanism by which the living cell is believed to synthesise proteins.

10. (*a*) Describe the properties of DNA that allow self-replication to take place.

(*b*) Why is it important that the replication of DNA should produce two exact copies of the original DNA molecule?

(*c*) What is meant by the statement that DNA replication is *semi-conservative*?

Describe the experimental evidence for this view.

11.

CH_2OH ... A

$$HO_2CCH_2CH_2CHCO_2H$$ with NH_2 B

C

(*a*) State the class of compounds to which each of the substances above belongs.

(*b*) Give the name of the type of polymer which is formed by each of the substances

(*c*) State the function of each of these polymers in the cell.

(*d*) Write formulae showing how 2 units of the monomers combine in each case.

6

VITAMINS

6.1 THE VITAMINS IN OUR DIET

Vitamins are needed in small amounts by the body.

The vitamins are a group of organic compounds which are needed in small amounts by the body. Vitamins are not a group of related compounds. They have a variety of formulae [see Figures 6.1A and 6.1B]. Vitamins are present in foods. The amounts are much smaller than those of proteins, lipids and carbohydrates, but the body needs much less. A varied diet supplies all the vitamins that a person needs.

As vitamins were discovered they were named according to the letters of the alphabet. In some cases it was found that a vitamin was really a mixture of two substances, and this is how names like Vitamin D_2 and Vitamin D_3 arose. As the study of vitamins continued, their chemical structures were worked out and many are now known by their chemical names.

Vitamin C and the eight members of the vitamin B group are soluble in water [see Figure 6.1A]. Vitamins A, D, E and K are fat-soluble: they dissolve in fats but not in water [see Figure 6.1B]. Table 6.1A lists the sources of vitamins in the diet.

FIGURE 6.1A
Some Water-soluble Vitamins

They are not a related group of compounds like lipids or nucleic acids. The formulae of some examples are shown.

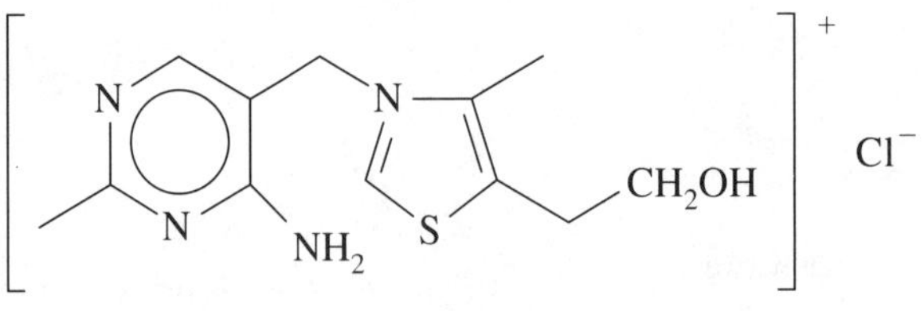

Thiamin (as the chloride) (vitamin B_1)

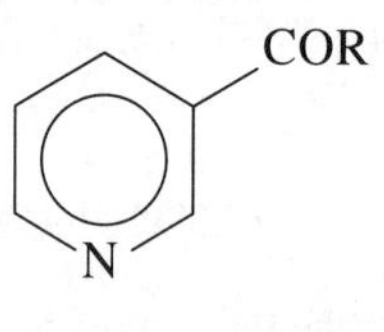

Nicotinic acid (niacin) (R = OH)
Nicotinamide (R = NH_2)

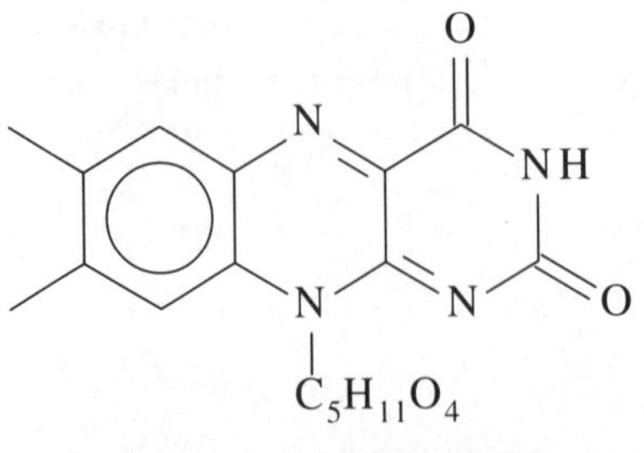

Riboflavin (vitamin B_2)

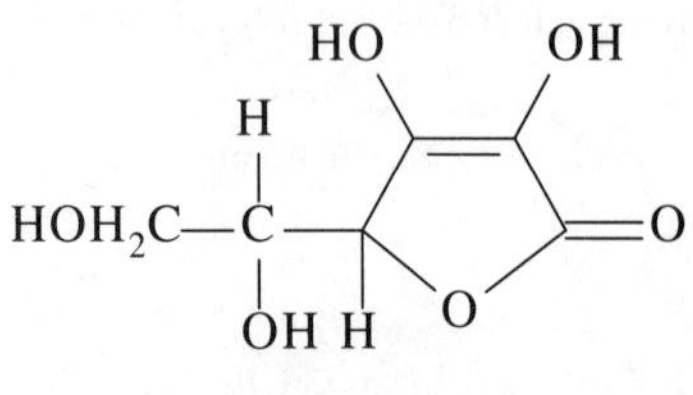

Ascorbic acid

82

FIGURE 6.1B
Some Fat-soluble
Vitamins

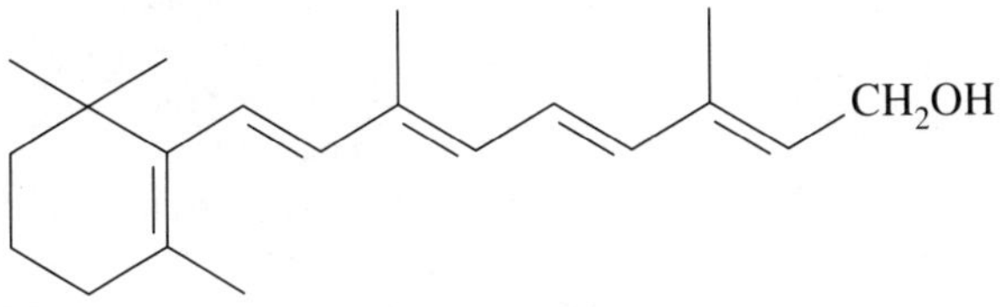

Retinol (vitamin A)

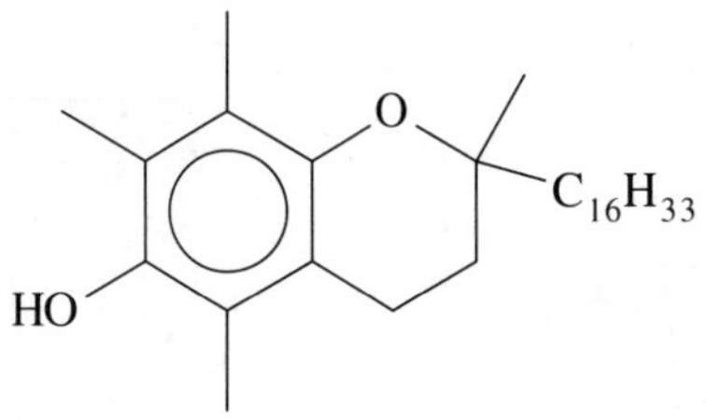

α-Tocopherol (one of several tocopherols
(vitamin E))

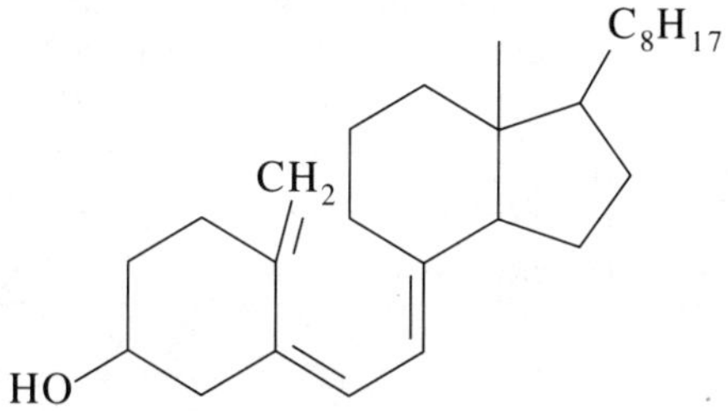

Cholecalciferol (vitamin D$_3$)

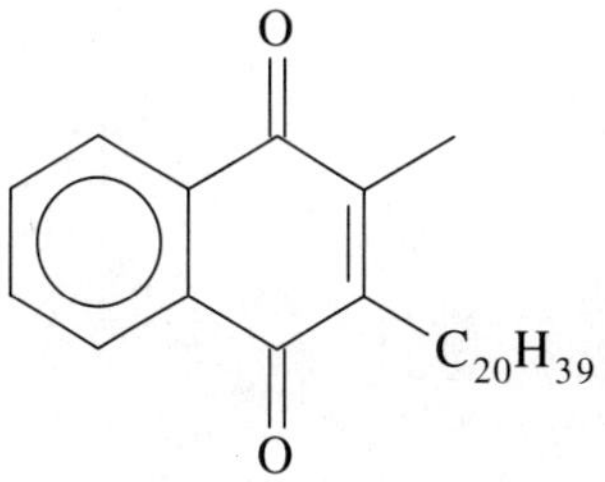

Vitamin K$_1$ (one of two vitamin K's)

*The dietary sources of
some vitamins are
tabulated.*

TABLE 6.1A
Dietary Sources of
Vitamins

Vitamin	*Source*	*Function*	*Requirement /mg*
A (retinol; β-carotene)	Fish liver oils, butter, margarine, carrots	Keeps eyes and skin healthy. Lack of vitamin A can cause blindness.	0.75
B$_1$ (thiamin)	Cereal germ, yeast	Assists growth, muscle function, nervous system, liberation of energy. Deficiency causes beriberi.	1.2
B$_2$ complex includes:		Similar to vitamin B$_1$.	
Riboflavin	Yeast, milk	Deficiency affects the eyes and causes mouth sores and cracks.	1.7
Nicotinic acid	Meat, yeast, wholemeal bread	Deficiency causes pellagra, a skin disease, and diarrhoea.	18
B$_{12}$	Yeast, animal protein	Formation of red blood cells	0.001

(continued)

Vitamin	Source	Function	Requirement /mg
C (ascorbic acid)	Fruits, vegetables	Necessary for the formation of blood and bone and for resistance to infection. Deficiency produces scurvy.	30
D (e.g. cholecalciferol)	Fish liver oil, egg yolk, butter, margarine	Necessary for healthy bones. Deficiency produces rickets.	0.0025
E (tocopherol)	Cereal germ, green leaves	Healthy reproduction, muscle health	—
K	Green leaves	Thought to resist blood clotting.	—

6.2 VITAMIN C

Vitamin C, ascorbic acid, is readily oxidised to dehydroascorbic acid.

Ascorbic acid

Vitamin C is ascorbic acid. It is easily oxidised by air, and it is water-soluble.

Ascorbic acid is oxidised by exposure to air. The oxidation is catalysed by oxidases which are present in the cells of foodstuffs. Since the rate of oxidation increases rapidly with the temperature and with the presence of alkali, the vitamin C content of a food is preserved better in a weakly acidic solution and by storage in the cold. During cooking, some ascorbic acid dissolves in the cooking water and some is oxidised. The volume of water used should be kept to a minimum. To minimise oxidation, fruits and vegetables should not be chopped or crushed before cooking as this sets free the oxidases which catalyse the oxidation of ascorbic acid. A good practice is to place vegetables in boiling water for cooking. Dissolved oxygen has been driven out of the water by boiling, and oxidases are deactivated at this temperature.

Methods of cooking fruits and vegetables should avoid chopping, alkaline conditions and prolonged boiling in water.

The addition of sodium hydrogencarbonate, which some cooks use to improve the colour of cooked green vegetables, should be avoided as alkali assists the oxidation of ascorbic acid. Microwave cookery avoids any loss of vitamin C through dissolution. Cooked food should be served as soon as possible to avoid destroying the ascorbic acid. [For cooking vegetables, see also § 11.1.]

6.2.1   DETERMINATION OF VITAMIN C CONTENT OF FRUIT JUICE

Ascorbic acid is a reducing agent. It is the only compound in food which can reduce 2,6-dichlorophenolindophenol (DCPIP). A pink compound is formed. This reaction can be used to test for vitamin C and to estimate it.

DCPIP is sold in the form of tablets. One tablet will oxidise 1.0 mg of ascorbic acid, but a solution of DCPIP must be standardised before use. A solution of DCPIP is prepared (4 tablets containing a total of 0.05 g of DCPIP in 100 cm^3 of solution). A standard solution of ascorbic acid is prepared (0.0500 g ascorbic acid in 250 cm^3 of a solution which contains phosphoric acid as anti-oxidant). The standard ascorbic acid solution is pipetted into a conical flask and titrated with DCPIP solution until a pink colouration is formed and persists for 15 seconds. From the titration, the mass of ascorbic acid which is oxidised by 1.00 cm^3 of DCPIP solution can be found.

The concentration of vitamin C in fruit juice can be found by titration against DCPIP.

A volume (50 cm^3) of fruit juice is pipetted into a conical flask, phosphoric acid is added to stabilise it, and the juice is titrated against DCPIP. From the titre, the mass of ascorbic acid in this volume of fruit juice can be calculated [see Questions on Chapter 6]. If the fruit juice contains a sulphite as a preservative, a little propanone is added before titration. This forms the hydrogensulphite compound of propanone and stops sulphite from interfering in the redox reaction.

6.3   VITAMIN SUPPLEMENTS

Vitamin supplements are added to some foods, e.g. flour and margarine.

Bread forms a substantial part of many people's daily diet. Since some vitamins are removed from wheat in the production of white flour, vitamin supplements are added to flour. They are vitamins B$_1$, B$_2$ and B$_3$.

Margarine must contain 8.5 mg of vitamin A and 55 μg of vitamin D per kilogram. To achieve these levels, vitamin supplements are added.

QUESTIONS ON CHAPTER 6

1. Joshi eats fish, vegetables and fruit but no meat. Which of the vitamins listed in Table 6.1A might be missing from his diet? What could he do to remedy any deficiency?

2. Enas never drinks milk. What vitamin listed in Table 6.1A might be missing from her diet?

3. Malindi is on a slimming diet and has cut out oil, butter, margarine and eggs. What vitamin could be deficient in her diet?

4. Gwynneth squeezed an orange and strained the juice into a 250 cm^3 volumetric flask. She added 50 cm^3 of dilute ethanoic acid as a preservative and made the volume up to 250 cm^3. She pipetted 10 cm^3 of the solution of orange juice into a conical flask. She titrated this with DCPIP solution, obtaining a titre of 30.0 cm^3. The DCPIP solution contained 4 tablets in 100 cm^3 of solution. Each tablet can oxidise 1.00 mg of ascorbic acid. How much vitamin C (mg) did the orange contain?

5. Describe how you could test a food sample for
(*a*) protein, (*b*) a reducing sugar, (*c*) a lipid, (*d*) vitamin C. [Refer to §§ 2, 3, 4 if necessary.]

7

MINERALS

7.1 MINERAL ELEMENTS AND TRACE ELEMENTS

The **mineral elements** are the elements, other than carbon, hydrogen, oxygen and nitrogen, which are required by the body. They make up about 4% by mass of the body. Some, e.g. calcium and iron, are present in relatively large amounts. Others are present in tiny amounts and are called **trace elements** [see Table 7.1A]. The elements are ingested as inorganic salts in food and as constituents of organic compounds, e.g. phosphorus and sulphur in proteins.

The elements such as calcium, iron and phosphorus that are required by the body are called mineral elements.

The major mineral elements and the trace elements are listed.

Major elements	Recommended daily intake for adult/mg
Calcium	500
Chlorine	—
Iron	10–12
Magnesium	200–300
Phosphorus	800
Potassium	800–1300
Sodium	—
Sulphur	—
Trace elements	*Probable daily requirement/mg*
Chromium	Trace
Cobalt	0.1
Copper	1–3
Fluorine	1–2
Iodine	0.06–0.15
Manganese	3–5
Molybdenum	Trace
Selenium	Trace
Zinc	10–20

TABLE 7.1A
Mineral Elements
Required by the Body

Of the elements tabulated, those which are sometimes deficient in the diet are calcium, iron, iodine and fluorine.

7.2 CALCIUM

Calcium is an essential element in the structure of bones and teeth. Some diets lack sufficient calcium, which is present in dairy products and other foods.

The body of an adult contains 1.0–1.5 kg of calcium. In addition to being an essential element in bones and teeth, calcium is required for blood clotting and for the functioning of muscles and nerves. There is an equilibrium between calcium ions in the blood and those in the skeleton. If the diet contains insufficient calcium, calcium passes out of the bones into the blood, thus weakening the skeleton. The absorption of calcium from food is assisted by vitamin D. Substances which interfere with the absorption of calcium are ethanedioic acid and phytic acid, which is present in wholemeal flour. These ligands combine with calcium ions to form complexes of high stability. Much of the calcium in the diet passes through the intestines as these complexes, which do not release calcium into the blood stream. Only half the calcium ingested in food is absorbed from the intestine. Sources of calcium in the diet include cheese, milk, sardines and white bread.

7.3 IRON

The body of an adult contains about 4 g of iron. Most of the iron is present in haemoglobin [see Figure 2.9A for the structure]. A lack of iron in the diet can therefore cause anaemia. All the cells of the body contain some iron, which takes part in enzyme-catalysed reactions. Iron is stored as ferritin, a protein–iron complex which is found in bone marrow, liver and spleen. If there is a shortage of iron in the diet, the number of red blood cells is reduced and the ability of the blood to transport oxygen is lowered, with a resulting lethargy and lack of energy.

Iron is a component of haemoglobin and therefore important for healthy red blood cells.

Some diets lack iron, which is present in red meats and some vegetables.

Of the iron ingested in food, only 5–20% is absorbed. A person who lacks iron can absorb more than someone with an adequate level of iron. Food contains iron in the iron(III) state, and this must be reduced to the iron(II) state before it can be absorbed. Vitamin C, being a reducing agent, helps in the reduction and therefore in the absorption of iron. Ethanedioic acid and phytic acid form insoluble complexes with iron and interfere with its absorption. Sources of iron in the diet include liver, kidney, other meats, lentils, wholemeal bread and vegetables. Iron compounds are added to white flour to bring the content up to 1.65 mg iron per 100 g of flour.

7.4 PHOSPHORUS

Phosphorus is an essential element in bones, teeth and ATP.

Phosphorus is an essential component of bones and teeth. As an element in adenosine triphosphate, ATP, it is involved in the release of energy to cells. Phosphorus occurs in most foods, especially cheese, sardines, eggs and brown bread. No reasonable diet lacks phosphorus.

7.5 SODIUM AND POTASSIUM

Potassium ions occur inside cells, and sodium ions occur in the fluid surrounding cells. Both sodium ions and potassium ions are essential in the regulation of the water content of the body and in the transmission of nerve impulses. The body requires

Sodium and potassium are essential in the regulation of the water content of the body and in the transmission of nerve impulses.

about 4 g of sodium chloride a day. An excess of salt is excreted or lost in sweat. In some kidney diseases, too much salt is retained by the body, and, in order to maintain the correct concentration of salt, water is retained also, and the result is oedema (excessive water in the tissues). Sources of sodium in the diet are yeast extract, bacon, cornflakes and salted butter. Sources of potassium are fruits, vegetables and many other foods.

7.6 IODINE

Iodine is needed for the synthesis of the hormone thyroxin.

Thyroxin is the hormone which is involved in the regulation of the oxidation of nutrients in cells. Iodine is needed for the synthesis of thyroxin. A deficiency of iodine causes enlargement of the thyroid gland, a condition called **goitre**. It is not uncommon in parts of the world where the soil and the diet lack iodide ion.

Sources of iodine in the diet are cereals, vegetables, milk and seafoods.

7.7 FLUORINE

Fluoride ion combines with the calcium hydroxide phosphate in tooth enamel to replace some of the hydroxide ion by fluoride ion.

$$Ca_5(PO_4)_3OH \rightarrow Ca_5(PO_4)_3(OH)_{1-x}F_x$$
Calcium hydroxide   Fluoridated calcium
phosphate   hydroxide phosphate

Fluorine, as fluoride ion, strengthens tooth enamel.

The treated enamel is more resistant to attack by acids and therefore helps to prevent tooth decay. Some Water Authorities add fluoride ion to drinking water to bring the level up to 1 ppm (part per million). The concentration of fluoride ion must not exceed 1.5 ppm because above this level it can cause brown stains on teeth. Sources of fluorine in the diet include seafoods and tea.

7.8 MAGNESIUM

Magnesium, as its ions, is an enzyme activator and is important in ATP synthesis, nerve function and muscle function.

The release of energy to cells is regulated by reactions which involve adding phosphate groups to ADP and removing phosphate groups from ATP [see §§ 9.2, 9.5]. Magnesium ions activate many of the enzymes that control these reactions. Magnesium ions are crucial in regulating nerve function and muscle contraction. They play a role in maintaining the structure of DNA [see § 5.2].

Magnesium occurs in green vegetables, nuts, cereals and seafoods. Magnesium deficiency is uncommon, but it is seen in alcoholics and victims of kwashiorkor [see § 15.1].

7.9 TOXIC ELEMENTS

There are toxic elements in foods.

Elements which are present in foods and which are not beneficial to life or are toxic are copper, arsenic, cadmium, mercury and lead.

7.9.1   LEAD

Among them is lead which can enter the water supply from lead pipes (which are being replaced) and from vehicle exhausts (of vehicles which do not use lead-free petrol).

Lead may be ingested in drinking water if the water has passed through lead pipes. These pipes were installed a century ago, but are still in use in some parts of the UK. Lead compounds from exhaust gases of motor vehicles can land on the surfaces of fruit and vegetables grown near roads. This source of contamination is reduced now that many vehicles use lead-free petrol. The effects of an intake of lead are a slowing down of mental processes and depression.

7.9.2   MERCURY

Mercury can enter the food chain by a number of routes:

● Seeds treated with mercury compounds as fungicides. The seeds are planted, crops grow, and rain washes mercury compounds into the soil.

● Waste water from factories which use mercury compounds is discharged into rivers.

● Mercury compounds which are used to treat timber are washed into the environment. Micro-organisms convert mercury salts into organo-mercury compounds, e.g. methylmercury salts and dimethylmercury salts. These compounds are ingested by marine organisms and freshwater organisms and become concentrated as they pass along a food chain which ends with man. The effects of mercury intake were seen at Minamata [see *ALC*, §18.8.2]. They include lack of muscular coordination and loss of mental powers.

Mercury is a toxic element which can enter the food chain from fungicides, waste water from factories, timber yards and other industries.

Despite the warning of Minamata, people are still exposed to mercury poisoning. Gold prospectors in Brazil, who concentrate gold dust by using mercury to form an amalgam with the gold, dispose of their used mercury into rivers, where it is concentrated up the food chain.

QUESTIONS ON CHAPTER 7

1. (*a*) What are the effects of a deficiency of calcium in the diet?

(*b*) How can this deficiency be rectified?

2. Flora is a vegetarian.

(*a*) Which element is most likely to be lacking in her diet if she is not careful?

(*b*) What type of foods will supply any deficiency of this element?

3. Molly hates milk. Which element may be lacking in her diet if she is not careful?
How can she make sure she gets enough of this element?

4. (*a*) What part does sodium chloride play in the regulation of human metabolism?

(*b*) What happens if the kidney fails to excrete excessive sodium chloride from the body?

5. (*a*) How can you tell whether a person has goitre?

(*b*) Why does goitre often attack members of a family? Does this indicate that the condition is genetically linked?

6. (*a*) Look at the formulae of calcium hydroxide phosphate and fluoridated calcium hydroxide phosphate. Suggest why fluoride protects teeth from decay.

(*b*) What is the effect of too much fluoride in the diet?

8

WATER

8.1 WATER FOR LIFE

All living organisms need water. It provides a medium in which nutrients, waste products and enzymes can dissolve and in which the biochemical reactions which maintain life can take place. Blood plasma is 90% water. In mammals it is the blood plasma that transports nutrients to and waste from cells in the body.

Water is the medium in which all the biochemical reactions in living organisms take place.

The body of an adult man contains about $40 \, dm^3$ of water (blood, tissue fluid and intra-cellular fluid). An adult man requires about $2.5 \, dm^3$ of water a day from drink, from food and from the oxidation in the body of glucose and other nutrients. A man excretes about $1.5 \, dm^3$ a day in urine and faeces; $0.6 \, dm^3$ through the skin and $0.3 \, dm^3$ in expired air.

8.2 WATER AS A SOLVENT

The polar character of its bonds makes water a good solvent for ionic compounds and for many organic compounds.

The ability of water to act as a solvent is due to its polar nature. It enables inorganic salts to dissolve by hydrating the ions, an exothermic process which provides the energy necessary to break up the crystal structure [see *ALC*, § 17.9]. Water is also a good solvent for compounds with which it can form hydrogen bonds, e.g. alcohols [see *ALC*, § 30.2.2], carboxylic acids [see *ALC*, § 33.3.1] and others.

8.3 COLLOIDAL DISPERSIONS

Many substances which have large molecules form colloidal dispersions in water. Many foods are colloidal dispersions, including emulsions, e.g. milk, margarine, mayonnaise.

When water is added to a substance which consists of very large molecules, instead of a solution a **colloidal dispersion** is formed. The substance with large molecules forms the **disperse phase**. and water forms the **dispersion medium** or the **continuous phase** [see *ALC*, § 9.5]. Examples of foods which are colloidal dispersions are beaten egg white (gas in liquid), bread (gas in solid), mayonnaise, milk and butter (liquid in liquid) and jam (liquid in solid). A dispersion of a liquid in a liquid is called an **emulsion**. Many food emulsions are emulsions of oil and water (e.g. milk, cream, salad cream, ice-cream) or water in oil (e.g. butter and margarine). To prevent oil and water separating, an **emulsifier** is added, a substance which contains a hydrophobic group which will bond to the oil and a hydrophilic group which will bond to water [see § 4.14 and *ALC*, § 33.13.3].

All foods contain water. The percentage varies from about 2% in biscuits and 35% in cheese to 70% in meat and 95% in lettuce. The water is available for enzyme activity and for the growth of micro-organisms. Both types of process make food deteriorate *All foods contain water.* and spoil [see § 12.4].

QUESTIONS ON CHAPTER 8

1. Explain the difference between a solution and a colloidal suspension.

2. In an emulsion, what is (*a*) the disperse phase and (*b*) the dispersion medium?

3. Give three examples of foods which are emulsions.

4. Explain (*a*) why water is a good solvent for monosaccharides and disaccharides (§ 3.2, 3.3) and (*b*) why water is a poor solvent for lipids [§ 4.1].

5. Explain how water is able to permeate cell surface membranes [see § 4.12].

9

METABOLIC PATHWAYS

9.1 THE NEED FOR CONTROL

The metabolism of an organism is the sum of the biochemical processes that take place within it; both catabolic processes and anabolic processes.

Metabolism is the term used for all the biochemical processes that take place within an organism. Some of these processes are **catabolic**: they involve the formation of substances with small molecules from substances with larger molecules. Cellular respiration is an example of **catabolism**. Some metabolic processes are **anabolic**: they involve the combination of substances with small molecules to form substances with larger molecules. Protein synthesis is an example of **anabolism**.

Even a simple organism, e.g. a bacterium, carries out hundreds of different chemical reactions. The reactions fall into a number of sequences, in which the product of one reaction is the substrate for the following reaction. Such a sequence of reactions, from starting materials to products, is called a **metabolic pathway**. Each step in a metabolic pathway is enzyme-catalysed. Enzymes are very effective catalysts, they operate only on specific substrates [see § 2.14], and they are sensitive to control by cofactors [see § 2.16].

A sequence of reactions, from starting materials to products, is called a metabolic pathway.

Many metabolic pathways begin with sugars, fatty acids and amino acids. In animals, these are derived from polysaccharides, fats and proteins in the diet. In plants, they are synthesised from simpler materials, e.g. carbon dioxide, water and molecular nitrogen, nitrate ion and ammonium ion. The first stage in the generation of energy is the breaking down of large molecules of food into smaller molecules of sugars, fatty acids, glycerol, amino acids, etc. In the second stage, these small molecules – fuel molecules – are converted into a small number of substances which play a crucial role in the liberation of energy. The most important of these is **acetyl coenzyme A (acetyl Co A)**.

The oxidation of carbohydrates is one of the chief sources of energy in living cells. When glucose is oxidised completely,

$$C_6H_{12}O_6(s) + 6O_2(g) \rightarrow 6CO_2(g) + 6H_2O(l); \Delta H^{\ominus}_{298} = -2820 \text{ kJ mol}^{-1}$$

The metabolic pathway that leads from glucose and oxygen to carbon dioxide and water is studied here. In the cell, the energy liberated from the oxidation of glucose is released gradually.

When this oxidation takes place in living cells, the energy is not all liberated immediately. If it were, the temperature of the cells would rise catastrophically and the cells would be killed. Some of the energy liberated may be used to maintain the body temperature of the organism, some is needed for doing mechanical work, e.g. in muscle cells, and some is used in the transport of substances into and out of cells. When the organism is growing, energy is required for the synthesis of cell components. In a mature organism, the chemical composition of the cells remains fairly constant, although many components are being used up and replaced. In this steady state, the cell needs a steady supply of energy.

9.2 GLYCOLYSIS

The first step in the metabolic pathway is glycolysis.

There is a metabolic pathway by which glucose is oxidised in a multi-stage process which releases energy gradually. The first stage [see Figure 9.2A] of the pathway converts glucose into pyruvate ion; this pathway is called **glycolysis**. Pyruvate ions are present in association with cations, e.g. Na^+. Pyruvate ion is converted in a **link reaction** into acetyl Co A. Glycolysis can be carried out in the absence of oxygen and releases only a fraction of the energy available from glucose. The second stage of the pathway involves the **tricarboxylic acid cycle**, and the third stage is the **electron transport chain** or **respiratory chain**. The metabolism of glucose can be summarised as:

$$\text{Glucose} \xrightarrow{\text{1. glycolysis}} \text{Pyruvate ion} \xrightarrow{\text{link reaction}} \text{Acetyl Co A} \xrightarrow[\text{3. electron transport chain}]{\text{2. tricarboxylic acid cycle}} \text{Carbon dioxide} + \text{Water}$$

The metabolic pathway involves **adenosine triphosphate (ATP)**.

Glucose is converted into pyruvate ion without the participation of oxygen. The interconversion of ADP + Pi and ATP is a vital part of the pathway. Each endothermic step is accompanied by the conversion of ATP into ADP. Each exothermic step is accompanied by the conversion of ADP into ATP.

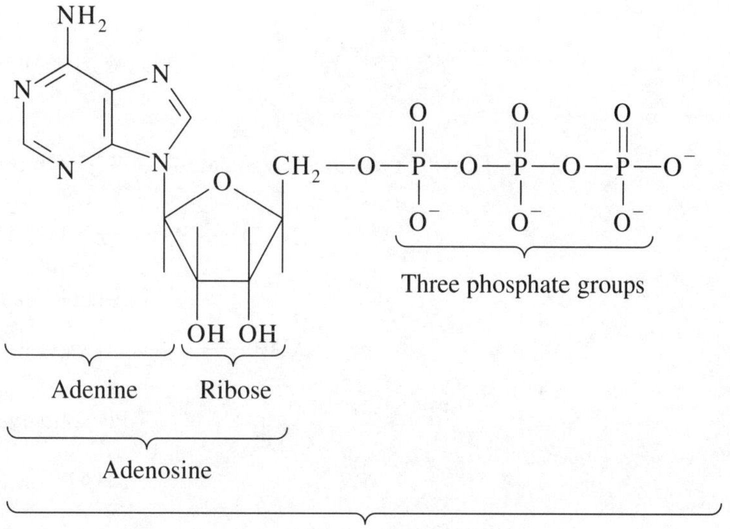

Adenosine triphosphate (ATP)

ATP is made in the cell by the addition of a phosphate ion to **adenosine diphosphate (ADP)**, a reaction called **phosphorylation**. Phosphorylation is endothermic; the hydrolysis of ATP is exothermic. Phosphorylation can be represented (writing Ad— for adenosine) as:

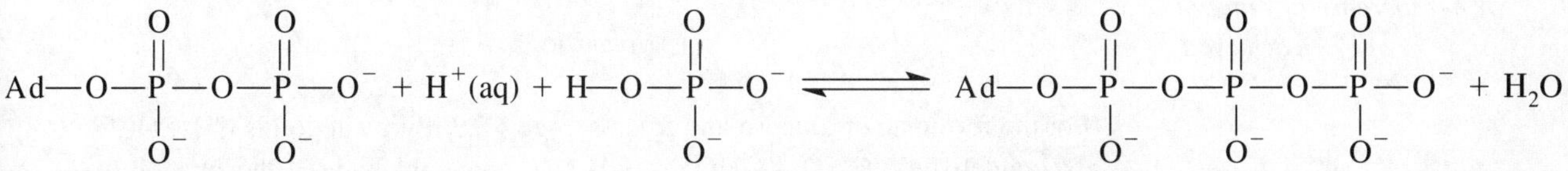

ADP + Hydrogen ion + Phosphate ion $\longrightarrow$ ATP + Water; $\Delta H^{\ominus}_{298} = 32 \text{ kJ mol}^{-1}$

ATP + Water $\longrightarrow$ ADP + Phosphate ion + Hydrogen ion; $\Delta H^{\ominus}_{298} = -32 \text{ kJ mol}^{-1}$

As we follow the stepwise oxidation of glucose, you will see how the endothermic reaction steps are coupled with the conversion of ATP into ADP, and the exothermic steps are coupled with the conversion of ADP into ATP. As a result, there is no sudden release of energy which would waste energy and also overheat the cells.

The steps in glycolysis, which converts glucose into pyruvate ion (2-oxopropanoate, $CH_3COCO_2^-$), are shown in Figure 9.2A. Glucose is converted into fructose-1,6-diphosphate and this is split into two trioses (3-carbon sugars) which are later converted into pyruvate ion (2-oxopropanoate, $CH_3COCO_2^-$). (You are not expected to remember details of this pathway.)

FIGURE 9.2A
Glycolysis: the
Conversion of Glucose
into Pyruvate ion

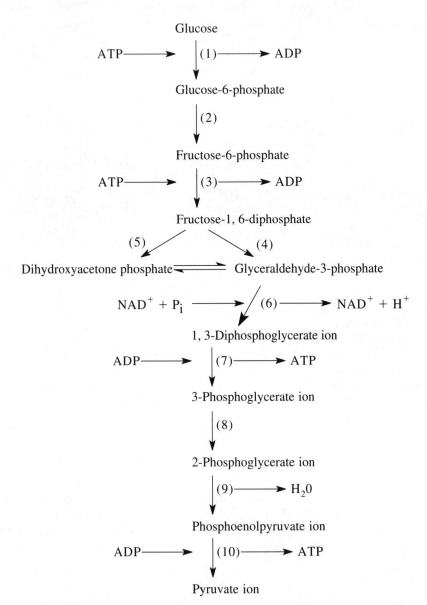

In glycolysis 1 mol of glucose is converted into 2 mol of pyruvate ion. No oxygen is used. A net increase of 2 mol of ATP is achieved.

(For the formulae of glucose and fructose, see § 3.2. Pyruvate ion is $CH_3COCO_2^-$, glyceraldehyde is $CHO—CHOH—CH_2OH$, glyceraldehyde-3-phosphate is $CHO—CHOH—CH_2OPO_3^{2-}$, and 3-phosphoglycerate ion is $HO_2C—CHOH—CH_2OPO_3^{2-}$. P_i = phosphate ion, HPO_4^{2-})

All the reactions are enzyme-catalysed.

Enzymes are involved at each stage of glycolysis. **Phosphotransferases** catalyse the transfer of phosphate groups. **Dehydrogenases** catalyse reactions in which hydrogen atoms are added to or removed from a coenzyme. **Isomerases** catalyse isomerisations. Glycolysis takes place in the cytosol [see § 1.1].

All the intermediates in glycolysis contain phosphate groups. The significance of this is twofold. Firstly, the phosphate group is required for converting ADP into ATP. Secondly, none of the phosphate intermediates nor ADP nor ATP can penetrate cell surface membranes. The cell surface membrane resists the passage of such highly charged substances [see §4.12].

Oxygen enters the metabolic pathway at a later stage.

The oxidising agent which oxidises fuels in the cells is molecular oxygen, O_2. By definition, as an oxidising agent, oxygen is an electron acceptor. The transfer of electrons from the fuel to oxygen is indirect. Electrons are transferred via an **electron transport chain** [see § 9.4]. The presence of a coenzyme to act as an electron acceptor is necessary.

Coenzymes are essential for the activity of many enzymes [see § 2.17]. Many coenzymes are derived from vitamins. Coenzymes act as hydrogen-carriers in the enzyme-catalysed oxidation of glucose. They are NAD^+ (nicotinamide adenine dinucleotide) and FAD (flavin adenine dinucleotide). The formula below shows the part of NAD^+ which acts as a hydrogen acceptor, and therefore as an electron acceptor, attached to R, which represents the rest of the molecule. The formula XH_2 represents the reducing agent. It gives one hydrogen atom to NAD^+ while the other becomes a hydrogen ion solvated by solvent molecules.

In glycolysis hydrogen atoms are removed by the coenzymes NAD^+ and FAD.
NAD^+ is reduced to $NADH + H^+$ and FAD is reduced to $FADH_2$.
NAD^+ and FAD are regenerated later in the pathway.

$$R-N^+\bigcirc-H \;+\; XH_2 \;\underset{}{\overset{enzyme}{\rightleftharpoons}}\; R-N\bigcirc\overset{H}{\underset{H}{}} \;+\; H^+ \;+\; X$$

$$\underset{CONH_2}{} \qquad\qquad\qquad\qquad \underset{CONH_2}{}$$

NAD$^+$ Reducing agent   NADH

or $NAD^+ + XH_2 \overset{enzyme}{\rightleftharpoons} NADH + H^+ + X$

In glycolysis, NAD^+ is the electron acceptor in Step 6 (the oxidation of glyceraldehyde-3-phosphate; see Figure 9.2A). NAD^+ is reduced to NADH; therefore NAD^+ must be regenerated by oxidation before glycolysis can proceed. In **aerobic oxidation**, the oxidation happens when NADH transfers electrons to oxygen via the electron transport chain. In **anaerobic conditions**, NAD^+ is oxidised by pyruvate ion, which is itself reduced to lactate ion [see §9.6].

The oxidation of NADH to NAD^+ and $FADH_2$ to FAD is linked to the production of ATP from ADP. The conversion of ADP plus phosphate ion into ATP is endothermic. The energy liberated in the enzyme-catalysed oxidation of food is used to enable the enzyme-catalysed conversion of ADP into ATP to take place. The energy is not transferred directly; it is transferred through the coenzyme reactions as shown below.

The energy released in the reduction of NAD^+ to $NADH + H^+$ is used in the conversion of $ADP + P_i$ into ATP.

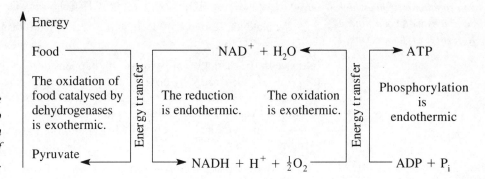

Glycolysis occurs along the Embden–Meyerhof pathway.

The glycolytic pathway is often called the **Embden–Meyerhof pathway** after two German scientists who worked out the details in the 1930s. The glycolysis pathway, together with the tricarboxylic acid cycle and the electron transport chain, provides a wonderful mechanism which enables the oxidation of glucose to be achieved at neutral pH, in dilute solution, at atmospheric pressure and at body temperature. It enables oxidation to proceed so that energy is released gradually and conserved by the conversion of ADP into ATP.

The rate of glycolysis is regulated by feedback control. This means inhibition of an enzyme by a product of the reaction. One of the enzymes in the glycolysis pathway is inhibited by ATP. ATP is one of the products of the reaction; therefore when the level of ATP rises the reaction slows down.

The rate at which glycolysis takes place is determined by the enzymes involved at different stages. The enzyme which catalyses the conversion of fructose-6-phosphate into fructose-1,6-diphosphate (Step 3 in Figure 9.2A) is inhibited by ATP and by citrate ion. Therefore when the level of ATP in the cell rises, the enzyme becomes less active, and the production of ATP falls. Fructose-6-phosphate accumulates, and this leads to the synthesis of glycogen. When the level of ATP in the cell falls, and the cell needs more ATP as a source of energy, the inhibition of the enzyme by ATP decreases, the activity of the enzyme increases and the rate of glycolysis increases. This type of control of a metabolic pathway is called **feedback control** [see §2.15]. Other enzymes in the pathway are inhibited by the products of the different stages. In each case, the more product is formed, the more the enzyme is inhibited and the more slowly the reaction proceeds. When the level of product drops, the enzyme activity increases and the rate of glycolysis increases.

CHECKPOINT 9.2

1. (*a*) Does a relatively low ATP/ADP ratio reflect a relatively high energy level in the cell or a relatively low one?

(*b*) Suggest how the cell will respond to a relatively low ATP/ADP ratio.

(*c*) How is the rate of glycolysis controlled?

2. Write the formula of pyruvic acid, which has the systematic name 2-oxopropanoic acid. How many moles of pyruvic acid are formed from one mole of glucose?

3. Why do cells employ a complicated mechanism for oxidising glucose instead of simply oxidising it directly to carbon dioxide and water?

9.3 THE TRICARBOXYLIC ACID CYCLE

Glycolysis is the first pathway in the oxidation of glucose to carbon dioxide and water. The oxidation is completed in the **tricarboxylic acid cycle**. Linking glycolysis with the tricarboxylic cycle is the conversion of pyruvate ion into **acetyl coenzyme A (acetyl Co A)**. This **link reaction** takes place in the mitochondria [see Figures 1.1A and B]. Pyruvate ion is the product of glycolysis. In the laboratory, the direct oxidation of pyruvate ion yields acetate ion (ethanoate ion) and carbon dioxide.

Glycolysis is followed by a link reaction. The link reaction oxidises pyruvate ion to ethanoate ion which combines with coenzyme A to form acetyl coenzyme A and carbon dioxide.

$$CH_3COCO_2^- + [O] \xrightarrow{\text{direct oxidation}} CH_3CO_2^- + CO_2$$

Pyruvate ion (2-oxopropanoate ion)   Acetate ion (ethanoate ion)

In the cells, the oxidation of pyruvate ion happens differently: it is brought about by the removal of hydrogen by NAD^+, rather than by the participation of oxygen. This is because the oxidation is strongly exothermic and must be coupled with endothermic reactions so that the energy released is not wasted. As soon as ethanoate ion (acetate ion) is produced in the cell from dehydrogenation of pyruvate, it combines with **coenzyme A (Co A)**, which is an adenine dinucleotide.

$$\text{Pyruvate ion} + \text{CoA} + NAD^+ \rightarrow \text{Acetyl Co A} + CO_2 + \text{NADH} + H^+$$

The conversion of one pyruvate ion into acetyl Co A removes one molecule of carbon dioxide and produces one molecule of reduced $NADH + H^+$. It remains to oxidise the acetyl (ethanoyl) groups to carbon dioxide and water. This takes place in a cycle of reactions which was discovered by Hans Krebs, a German-born British scientist, in the 1930s. Krebs and a Hungarian worker called Albert Szent-Györgyi had observed that dicarboxylic acids and tricarboxylic acids have powerful stimulating effects on the oxidation of glucose in muscle tissue. Krebs formulated a cycle in which organic acids of these types take part in a metabolic sequence. The last intermediate in the sequence, oxaloacetate (2-oxobutane-1,4-dioate) reacts with acetyl Co A to make citrate ion (2-hydroxypropane-1,2,3-tricarboxylate), which is the first intermediate and can start the cycle over again. A summary of the **tricarboxylic acid cycle**, which is often called the **Krebs cycle** and also the **citric acid cycle**, is shown in Figure 9.3A (GDP is guanidine diphosphate). The anions are present in a medium which contains cations, e.g. Na^+, to balance the charges. You are not expected to remember details of this cycle.

FIGURE 9.3A
Summary of the
Tricarboxylic Acid Cycle
(Krebs Cycle)

In the second stage of the pathway, acetyl Co A is oxidised to carbon dioxide and water. This happens in the tricarboxylic acid (TCA) cycle, also called the Krebs cycle and the citric acid cycle.

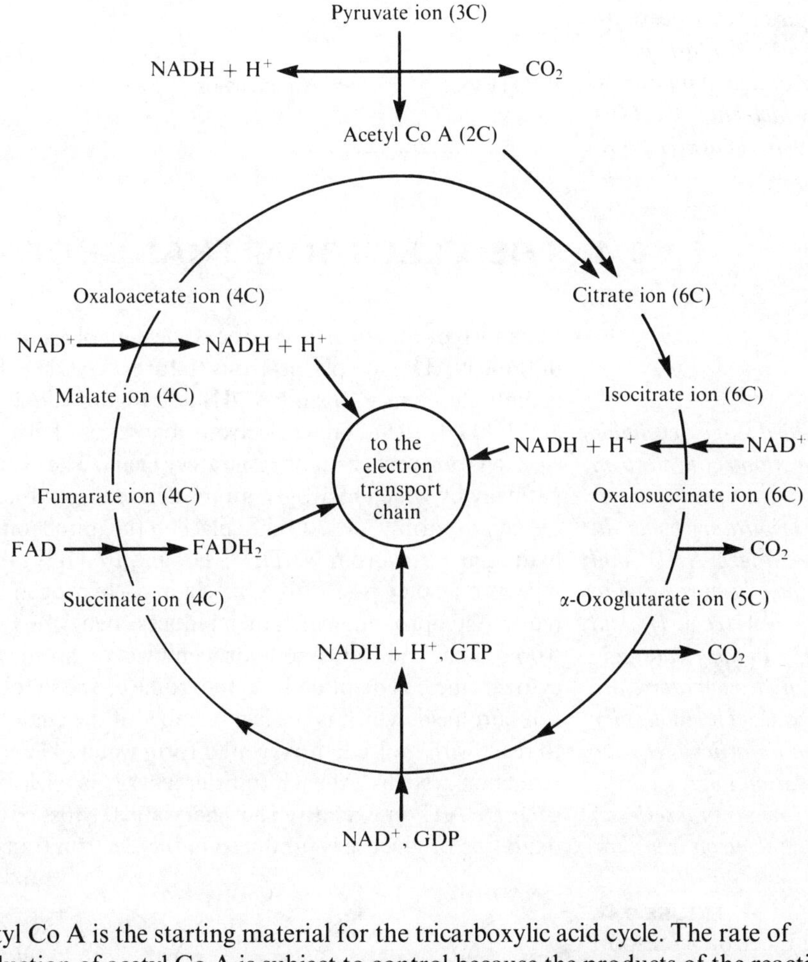

Acetyl Co A enters the cycle and carbon dioxide is evolved while hydrogen passes to the electron transport chain.

Acetyl Co A is the starting material for the tricarboxylic acid cycle. The rate of production of acetyl Co A is subject to control because the products of the reaction, acetyl Co A and NADH, both inhibit the enzyme which catalyses the reaction.

In the tricarboxylic acid cycle (the citric acid cycle), the 2-carbon group in acetyl Co A combines with the anion of the 4-carbon acid oxaloacetic acid to form citrate ion (with 6 carbon atoms). Citrate ion is then oxidised to the anion of a 5-carbon acid and carbon dioxide; then to the anion of a 4-carbon acid, succinic acid, and carbon dioxide. Succinate ion is converted in steps into oxaloacetate ion, and the cycle begins over again.

The TCA cycle allows ethanoate to be oxidised in stages and release its energy in small steps.

The useful aspect of the tricarboxylic acid cycle is that it provides a mechanism for acetate (ethanoate) ion to be oxidised in stages and release its energy in small steps.

The breakdown of fatty acids and amino acids yields products which can enter the cycle and be oxidised [see Figure 9.3B]

FIGURE 9.3B
Carbohydrates, Lipids and Proteins in Relation to the Tricarboxylic Acid Cycle
---

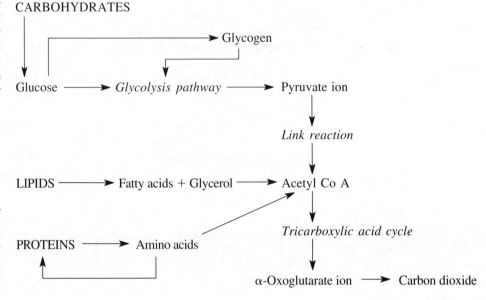

CARBOHYDRATES

Lipids and proteins are digested to fatty acids, glycerol and amino acids, all of which enter the TCA cycle via acetyl Co A.

9.4 THE ELECTRON TRANSPORT CHAIN

The third stage in cellular respiration is the electron transport chain or respiratory chain. In earlier stages, NAD⁺ and FAD have been reduced to NADH + H⁺ and FADH₂. In this final stage, the hydrogen atoms are oxidised by molecular oxygen to form water. The oxidation takes place by means of a series of electron transfers.

Each turn of the tricarboxylic acid cycle involves four dehydrogenation steps. In three of these NAD^+ accepts electrons to form $NADH + H^+$, and in the fourth FAD accepts electrons to form $FADH_2$ [see Figure 9.3A]. The reduced coenzymes NADH and $FADH_2$ then donate electrons to a series of electron carriers that form the **electron transport chain** or **respiratory chain**. The electron transport chain is the final pathway by which hydrogen atoms derived from fuel molecules are oxidised by oxygen to form water. It takes place in the mitochondria [see Figure 1.1A and B]. The hydrogen atoms from $NADH + H^+$ and $FADH_2$ take part in a series of redox reactions as they pass through a chain of 'hydrogen carriers' [see Figure 9.4A]. They reduce flavoprotein, which then reduces coenzyme Q (CoQ) to $CoQH_2$. This is oxidised back to CoQ and hydrogen by cytochrome b. By means of electron transfers, cytochrome b is oxidised by cytochrome c, and cytochrome c is oxidised by cytochrome a, which is the last member of the chain and allows the hydrogen atoms to react with molecular oxygen to form water. The various carriers are oxidation–reduction systems. At each transfer, energy is released. Some of this energy is used to generate ATP in **oxidative phosphorylation**. Most of the ATP formed through the oxidation of glucose is produced in the electron transport chain.

FIGURE 9.4A
The Electron Transport Chain (Respiratory Chain)

Note

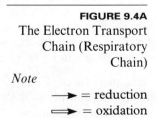
⟶ = reduction
⟹ = oxidation

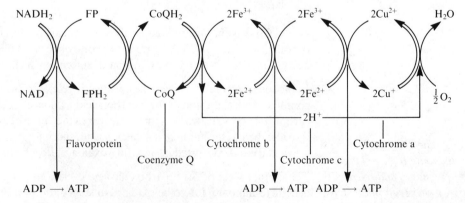

Flavoprotein is a coenzyme derived from vitamin B_2. Coenzyme Q has a 6-carbon ring structure. Cytochromes are proteins of relatively low molar masses with haem as a prosthetic group.

1. A summary of the Krebs cycle is shown in the figure.

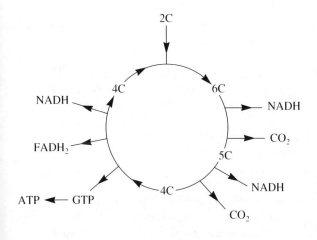

(a) Name the 2-carbon intermediate and the 6-carbon intermediate.

(b) Say how many moles of reduced coenzyme are formed in each turn of the cycle.

(c) What other names are given to this cycle?

2. (a) What processes are joined by the **link reaction**?

(b) Name the reactants and the products.

(c) What coenzymes must be present? What function do they serve?

3. What part is played in aerobic respiration by (a) acetyl coenzyme A, (b) the electron transport chain?

9.5 ATP

Many different reactions can provide the energy needed for the synthesis of ATP. Many different reactions draw on the energy released in the conversion of ATP into ADP. ATP is needed during the breakdown of glucose, fatty acids and amino acids, in the synthesis of macromolecules, e.g. nucleotides, in the contraction of muscles, in the transmission of nerve impulses, in the transport of substances into and out of cells. In the absence of ATP, the cell runs down, ceases to be able to perform its functions, and begins to use up all its remaining food reserves.

FIGURE 9.5A
ATP, ADP and Energy

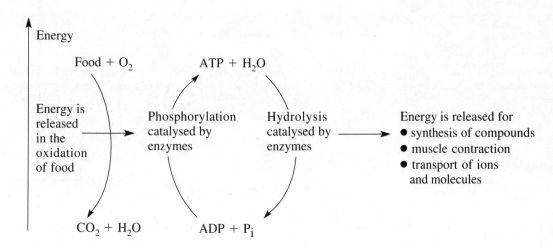

The energy released in the conversion of ATP into ADP + P_i is used to activate all the reactions which take place in living cells.

The energy stored in ATP is made available to processes within the cell only when it is needed and a specific enzyme catalyses its hydrolysis to provide energy for a specific process. In this way, ATP acts as an energy 'bank'; the formation of ATP from ADP takes place when energy is released through the oxidation of glucose, and the hydrolysis of ATP to ADP takes place when the cells need energy for a specific cell process [see Figure 9.5A; P_i = phosphate]. When the level of ATP is high, the cell slows down on the oxidation of glucose and builds up storage reserves of fats and carbohydrates.

How many moles of ATP are produced when one mole of glucose is oxidised via this pathway [see Figure 9.5B]?

Glycolysis gives 2 moles of ATP per mole of glucose. The tricarboxylic acid cycle gives 2 moles of ATP per mole of glucose. The respiratory chain gives 34 moles of ATP per mole of glucose. The total is 38 moles of ATP per mole of glucose.

When 1 mol of glucose is oxidised via this pathway, 38 mol of ATP are produced.

The energy released in the conversion ATP → ADP + P_i = 31 kJ mol^{-1}
Therefore energy released by oxidation of glucose = 38×31 = 1178 kJ mol^{-1}
The oxidation of glucose directly to carbon dioxide and water gives 2820 kJ mol^{-1}

The supply of energy via ATP therefore represents an efficiency of 42%. The remainder of the energy of the chemical bonds in glucose is converted into heat. The overall efficiency is less than this because there is a loss of energy in the transport of substances to the cycle and the transport of substances away from the cycle. Although the overall efficiency is less than 42%, it compares well with coal-fired and oil-fired power stations, which run at an efficiency of 21–25%, and motor vehicles which have an efficiency of about 20%.

FIGURE 9.5B
ATP Production

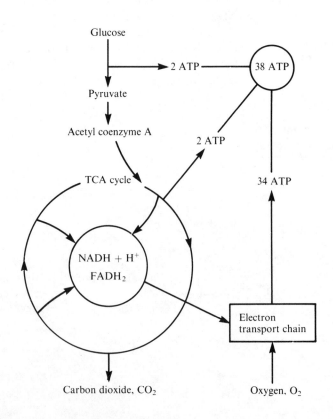

9.6 ANAEROBIC PATHWAYS

The aerobic pathway ends when glucose is completely oxidised to carbon dioxide and water, when NAD^+ and FAD have been regenerated and all the ATP (38 moles per mole of glucose oxidised) has been synthesised. When muscles are contracting very rapidly, for example when a mammal is running, there may be insufficient oxygen for complete oxidation. Then **anaerobic oxidation** occurs and lactate ion is formed.

When there is insufficient oxygen for complete oxidation, anaerobic oxidation takes over.

$$CH_3COCO_2^- + NADH + H^+ \rightarrow CH_3CHOHCO_2^- + NAD^+$$

Pyruvate ion   Lactate ion

This anaerobic pathway ends in lactate ion, with the synthesis of 2 moles of ATP per mole of glucose oxidised. Lactate ion can be further oxidised if oxygen becomes available later, with the release of more energy. Lactate ion fermentation is common in mammals because it enables them to withstand short periods of oxygen deficiency, e.g. during very strenuous exercise. Lactate ion fermentation cannot continue indefinitely: as lactate ion accumulates, it causes cramp, which eventually stops the muscles form working.

Pyruvate ion is converted into lactate ion. An oxygen debt is built up.

As lactate ion is fermented the mammal builds up an **oxygen debt**. This is repaid as soon as possible. You have seen athletes panting for breath when they have just completed a race. The rapid deep breaths which they inhale are used to oxidise the lactate ion.

The energy released is less than that in aerobic respiration.

In the aerobic respiration of glucose to carbon dioxide and later, the energy released is given by

$$C_6H_{12}O_6 + 6O_2 \rightarrow 6CO_2 + 6H_2O; \quad \Delta H^{\ominus} = -2880 \, kJ \, mol^{-1}$$

FIGURE 9.6A
Summary of the Cellular
Respiration of Glucose

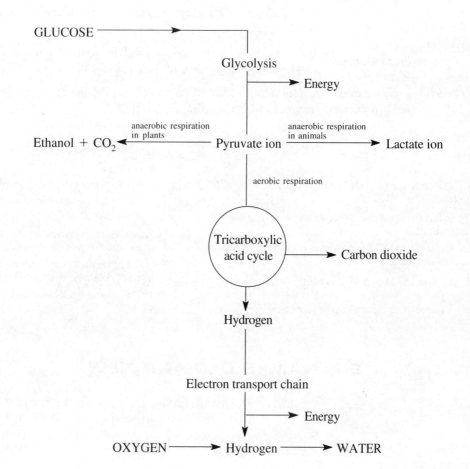

In the anaerobic respiration of glucose to lactic acid, the energy released is given by

$$C_6H_{12}O_6 \rightarrow 2CH_3CHOHCO_2H; \quad \Delta H^{\ominus} = -150 \, kJ \, mol^{-1}$$

In some micro-organisms, NAD^+ needed for glycolysis is regenerated by the reduction of pyruvic acid to ethanol and carbon dioxide; this process is known as **fermentation**.

$$CH_3COCO_2H + NADH + H^+ \rightarrow C_2H_5OH + CO_2 + NAD^+$$

Pyruvic acid   Ethanol   Carbon dioxide

Glucose can also be respired anaerobically to ethanol and carbon dioxide; this is fermentation. It releases less energy than aerobic respiration. Fermentation is catalysed by enzymes in yeast. It is important in brewing and baking.

In the overall anaerobic respiration of glucose to ethanol the energy released is given by

$$C_6H_{12}O_6 \rightarrow 2C_2H_5OH + 2CO_2; \quad \Delta H^{\ominus} = -210 \, kJ \, mol^{-1}$$

Alcoholic fermentation is catalysed by enzymes in yeast. The ethanol formed accumulates in the aqueous medium surrounding yeast until the concentration becomes high enough to kill the yeast. Alcoholic fermentation is of great economic importance in the brewing industry, where ethanol is the valuable product, and in the baking industry, where carbon dioxide is the more important product [see § 11.3].

9.6.1   METABOLIC CONTROL

The control of metabolic pathways takes place according to the following general principles:

1. The rate-controlling step is usually the first step of a pathway or the first irreversible step.

2. The activity of the enzyme which controls this step can be altered by activators and by inhibitors.

The control of metabolic pathways follows a number of general principles which are listed here.

3. The synthesis of the enzyme which catalyses the rate-determining step may be suppressed by the product of the metabolic pathway.

4. Hormones which act on the cell membranes of target cells affect the activity of enzymes within the cells.

CHECKPOINT 9.6

1. (*a*) Name the three chemical groups within the ATP molecule.

(*b*) What change in the molecule takes place in cellular respiration?

(*c*) Explain why ATP is viewed as occupying an essential position in metabolism.

2. (*a*) What yield of ATP is obtained from 1 mol of glucose
 (i) by aerobic metabolism to carbon dioxide and water,
(ii) by anaerobic metabolism to lactic acid?

(*b*) Why is more ATP generated in (i) than in (ii)?

(*c*) Why does anaerobic respiration produce less energy than aerobic respiration?

3. (*a*) State one reaction that occurs during the fermentation of glucose. How is it catalysed?

(*b*) What is the cause of the cramps which athletes suffer during prolonged strenuous exercise?

(*c*) Why does the ethanol content of wine not rise above 14%?

9.7   NAMES OLD AND NEW

Biochemists still use the traditional names for the intermediates in glycolysis and the TCA cycle. This is confusing for chemists who use the IUPAC systematic names. Some of the compounds in question are listed under their traditional names and their systematic names in Table 9.7A.

Traditional name	Systematic name
Acetic acid	Ethanoic acid
Citric acid	2-Hydroxypropane-1,2,3-tricarboxylic acid
Glyceraldehyde	2,3-Dihydroxypropanal
Glycerol	Propane-1,2,3-triol
Lactic acid	2-Hydroxypropanoic acid
Oxaloacetic acid	2-Oxobutane-1,4-dioic acid
Oxalosuccinic acid	2-Oxopentane-1,3,5-trioic acid
α-Oxoglutaric acid	2-Oxopentane-1,5-dioic acid
Pyruvic acid	2-Oxopropanoic acid
Succinic acid	Butane-1,4-dioic acid

Traditional names and systematic names of some compounds are listed.

TABLE 9.7A

9.8 HORMONES

The activities of cells are controlled by the nervous system and the endocrine system. The endocrine glands secrete hormones.

Enzymes control the flow of substances along metabolic pathways within the cells. In higher organisms, which consist of huge numbers of cells of different types, there must be a mechanism of coordinating the activities of different cells. Communication between the cells, tissues and organs of a mammal involves the **nervous system** and the **endocrine system**. A **gland** is a structure which secretes a certain chemical. There are two types of gland: **exocrine glands** and **endocrine glands**. In exocrine glands, the secretion is delivered along a duct to the site of action. In endocrine glands, also called **ductless glands**, the secretion is passed directly into the organism's fluids, e.g. blood. The nervous system and the endocrine system work in a coordinated manner to regulate the functions of the body. The combination of the two systems enables communication between cells to take place and results in the release of chemicals in the cells. Plant hormones are often called **plant growth substances**.

Hormones pass in the bloodstream to reach their target organ or tissue. They are called 'chemical messengers'. On arrival at its target the hormone binds to a receptor cell on the surface or within a target cell.

The endocrine glands secrete **hormones**. Produced in one part of the body, they enter the bloodstream and pass to a tissue or organ where they exercise a specific effect which regulates the action of that organ or tissue. Hormones are compounds of different types: proteins, peptides, amines, fatty acids and steroids. Hormones are described as **chemical messengers**. Each hormone has a '**target organ or tissue**' [see Table 9.8]. When it reaches its target, the hormone binds to certain receptors on the surface or within certain '**target cells**'. There are different ways in which the hormone modifies the metabolism of the target cells:

1. A hormone may alter the permeability of the cell surface membrane. This is how insulin increases the uptake of glucose in cells.

2. A hormone may bind to a receptor site in a cell surface membrane and cause the release of a second messenger inside the cell; see adrenalin (below).

3. Many hormones inhibit or activate enzymes. This is how thyroxine works to control the electron transport chain.

Hormones exercise their effects by a number of methods, which are listed.

4. Steroid hormones pass through the cell surface membrane and the nuclear envelope and into the nucleus where they activate genes and stimulate the formation of messenger RNA.

Some human hormones and their functions are listed.

Gland or tissue	Hormone	Function of hormone
Thyroid	Thyroxine	Regulates growth and development
Parathyroid	Parathormone	Regulates level of calcium and phosphate in blood
Islets of Langerhans	Insulin	Lowers blood glucose level
Pancreas	Glucagon	Raises blood glucose level
Adrenal cortex	Glucocorticoids	Control breakdown of proteins and synthesis of glycogen
Adrenal medulla	Adrenalin	Stimulates body to prepare for 'fight or flight'
Ovary	Oestrogens	Influence development of female characteristics and sex organs
	Progesterone	Inhibits ovulation during pregnancy
Testis	Testosterone	Influences development of male characteristics and sex organs

TABLE 9.8
Some Human Hormones

The secretion of a particular hormone by an endocrine gland may be triggered by the nervous system. The hormone adrenalin is the 'fight or flight' hormone. It is released from the cells of the adrenal medulla by the arrival of nerve impulses when an animal feels anxiety or stress or is in danger. Adrenalin binds to receptor sites in cell surface membranes and causes the release inside the cells of a **second messenger** which triggers a sequence of enzyme-catalysed reactions. These reactions produce a response to the original nerve impulse; they increase the rate of heartbeat, increase blood pressure, increase air flow to the lungs, stimulate the conversion of glycogen into glucose and increase mental awareness. These responses assist an animal to utilise its reserves of strength either to fight an aggressor or to flee from danger or to accomplish a feat of strength which would be beyond it without the help of adrenalin.

The mode of action of the hormone adrenalin is to bind to a cell surface membrane and cause the release inside the cell of a second messenger.

9.8.1 HORMONES IN MEDICINE

Alternatively, the trigger for hormone action may be the concentration of a certain compound in the tissues. The hormone **insulin** is released in response to an increase in the level of glucose in the blood. Insulin is a **polypeptide hormone** which is secreted by the islets of Langerhans in the pancreas and circulated in the bloodstream. The cells of the brain and nervous system can use only glucose as their energy supply. The blood glucose level is therefore critical for the normal functioning of the brain. The normal level of glucose is 60–100 mg per 100 cm^3 of blood. Immediately after a meal, the rise in the blood sugar level makes the pancreas secrete insulin. Insulin speeds up the passage of glucose into the liver, muscle and fatty tissue cells. In the liver it stimulates glycolysis, fatty acid synthesis and glycogen formation. As time passes and the glucose in the blood is metabolised, the blood sugar level falls. This causes the pancreas to stop producing insulin and to start producing another hormone, **glucagon**. This has the opposite effect to insulin: it stimulates the liver to convert glycogen into glucose. The release of glucose maintains the blood glucose at its normal level. Insulin and other peptide hormones become attached to specific receptor sites on the surface of target cell membranes, triggering changes within the cells.

The hormone insulin stimulates the formation of glycogen. The hormone glucagon stimulates the conversion of glycogen into glucose. Between them, the two hormones maintain the level of glucose in the blood at its normal level.

Diabetics do not manufacture enough insulin.

Diabetics do not manufacture enough insulin, and glucose is excreted in their urine. They inject insulin daily: it cannot be ingested orally because, being a polypeptide, it is hydrolysed by proteolytic enzymes in the gut. **Hyperglycaemia** is a very high blood sugar level. It causes the excretion of glucose in the urine and indicates diabetes or kidney failure. **Hypoglycaemia** is a very low blood sugar level. The brain becomes starved of glucose, and dizziness and fainting may occur. In severe cases, convulsions, shock and coma may follow. Hypoglycaemia may be caused by the over-injection of insulin and is called **insulin shock**. [For synthetic insulin, see § 5.8.]

Progesterone is a female hormone. Hormones similar to progesterone are used as synthetic oral contraceptives.

Progesterone is the hormone which suppresses ovulation during pregnancy. It produces little effect when taken orally because it is affected by digestive enzymes. A large number of synthetic compounds which are similar in structure to progesterone can also prevent the release of ova while resisting digestion. These synthetic hormones are manufactured in large quantities and form the basis of a number of **oral contraceptives**.

9.8.2 HORMONES IN MEAT PRODUCTION

Steroid hormones, e.g. oestrogens and testosterone, act differently from peptide hormones. A steroid hormone enters a cell and binds to a receptor site on a large protein molecule. The steroid-receptor group is transferred to the nucleus of the cell, where it activates specific genes. Steroid hormones control the development of sex organs and secondary male and female characteristics.

Steroid hormones, both male and female, are used to promote muscle development in livestock.

Livestock show increased growth rates when both male and female sex hormones are present in the blood stream. Maximum meat growth can be achieved by treating male animals with female sex hormones and female animals with male sex hormones. The hormones are either supplied in the diet or implanted in pellets under the skin.

Synthetic anabolic steroids are used for the same purpose.

Anabolic hormones are used in meat production. They stimulate the deposition of protein in animals. They are similar in structure to male and female sex hormones. The use of hormones to improve meat production must be carefully controlled because anabolic agents in the meat could have serious effects on the consumer. Naturally occurring sex hormones are destroyed during digestion and therefore do not pose a risk. Synthetic anabolic compounds, on the other hand, are resistant to digestion and there is therefore a question mark over their use. Meat samples are regularly tested for the presence of anabolic agents.

9.8.3 HORMONES IN INSECT CONTROL

There are problems over the use of insecticides for pest control. Many of the insecticides used are non-biodegradable. Others present a hazard to the people who have to apply them [see *ALC*, § 32.11]. The hormone control of insects is used in a number of cases as a substitute for insecticides. A substance called **juvenile hormone** controls the development of insects as they pass through several larval stages. If the insect is to develop into a mature adult, the hormone supply must be cut off after a certain stage. If an external supply of juvenile hormone is provided, an insect continues to develop as a larva or changes into an immature adult which cannot reproduce. The end result has several advantages over using an insecticide:

● The hormone is specific; it does not kill beneficial soil organisms.

● The insect is unlikely to develop a resistance to juvenile hormone because it needs the hormone at some stages of its development.

Hormones are used in insect control to inhibit reproduction.

● Since juvenile hormone is a natural substance, it is biodegradable and non-polluting.

9.8.4   HORMONES IN PLANT CULTIVATION

Plant hormones or **plant growth substances** called **auxins**, control the rate of metabolism and growth in plants, the ripening of fruit and the germination of seeds. An example is indole-3-ethanoic acid (known as **IAA** which stands for its old name, indole-3-acetic acid) which promotes plant growth.

Plant hormones – plant growth substances – are used to promote plant growth and to increase fruit crops.

—CH$_2$CO$_2$H   —CH$_2$CH$_2$CH$_2$CO$_2$H

Indole-3-ethanoic acid (IAA)   Indole-3-butanoic acid (IBA)

Plants synthesise IAA and also contain an oxidase enzyme which inactivates IAA. The balance between synthesis and inactivation keeps the IAA concentration at a low level. Horticulturists treat plants with synthetic compounds which have the activity of auxins but are not inactivated by the oxidase. The treatment may encourage the development of fruit or prevent fruit from dropping before it is harvested. The **synthetic hormone**, indole-3-butanoic acid (**IBA**) is used to promote the rooting of plant cuttings. It has advantages over IAA: it is oxidised only slowly and it is not easily transported through the plant so it does not promote growth in other regions.

The **synthetic hormone** 2,4-D (2,4-dichlorophenoxyethanoic acid) acts as a herbicide. When administered to plants, it produces such rapid growth that many species cannot survive. Weeds, which are broad-leaved plants, are killed while narrow-leaved grass and cereals survive.

Some synthetic plant growth substances stimulate growth to such an extent that the plant does not survive; these are used as weedkillers.

OCH$_2$CO$_2$H   OCH$_2$CO$_2$H

2,4-Dichlorophenoxyethanoic acid (2,4-D)   3,5-Dichlorophenoxyethanoic acid (3,5-D)

The compound 3,5-dichlorophenoxyethanoic acid (3,5-D) is similar in structure to auxins but does not have a hormonal action. It acts in the reverse sense, producing dwarf plants. It probably prevents the binding of natural auxins to the hormone receptor site. The dwarf plants may have advantages: they are strong and healthy, able to withstand high winds and easy to harvest.

CHECKPOINT 9.8

1. (*a*) Explain what a hormone is.

(*b*) State how hormones are able to bring about changes in cells.

(*c*) Name three mammalian hormones and say what function each performs.

2. Outline the importance of the hormone insulin in exerting control over metabolic processes.

3. Describe the use of a plant growth substance in plant cultivation.

QUESTIONS ON CHAPTER 9

1.

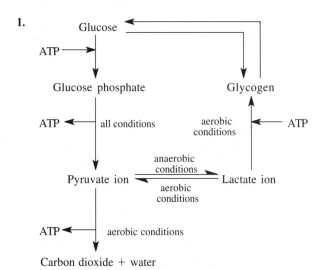

The flow chart summarises the processes that are involved in the release of energy in the muscles during and after exercise. Use it to help you to answer the following questions.

(*a*) Explain why sprinters have more lactate ion in their blood at the end of a race than at the beginning.

(*b*) Why does a sprinter's oxygen consumption remain high for a time after the race is over?

(*c*) After the race, some lactate ion is oxidised with the formation of ATP, and some lactate ion is converted into glycogen. Explain why the second course is an efficient use of lactate ion.

(*d*) Why is it an advantage to respire glycogen, rather than glucose, during strenuous exercise?

(*e*) Mention one way in which training improves an athlete's ability to gain energy by respiration.

2. Explain the meaning of 'cellular respiration'. Describe the part played by each of the following in cell respiration:

(*a*) glycogen, (*b*) fermentation, (*c*) oxidative phosphorylation, (*d*) electron transport system.

3. Why can insulin not be given to diabetics in the form of a pill?

4. The oxidation of glucose by a cell can be divided into two stages: an aerobic stage and an anaerobic stage.

(*a*) What name is given to the anaerobic stage?

(*b*) Write the equation for the overall reaction that takes place in this stage.

(*c*) Where in the cell does the oxidation of glucose occur?

(*d*) Which stage produces more energy: anaerobic or aerobic?

(*e*) Name two series of reactions that occur in the aerobic stage and give a general description of the reactions which occur in each.

(*f*) Write the equation for the overall reaction that occurs in the aerobic stage.

5. Jane's alarm clock fails to go off, and she wakes up late for an exam. She jumps out of bed, throws on some clothes, runs all the way to college and races up three flights of stairs to the exam hall. She collapses in her seat, gasping for breath, with her legs feeling like rubber.
Describe the events that have happened in Jane's muscles during this experience. Why is she gasping for breath? Why are her muscles so weak?

6. Ahmad's brain requires a constant supply of glucose. He eats only three times a day.
Explain how Ahmad's body maintains a fairly constant blood sugar level even though he eats only a few times a day.

7. (*a*) What is the key role which ATP plays in metabolism?

(*b*) Describe two cellular processes in which ATP is involved.

(*c*) Name a substance which is synthesised from the deoxyribose derivative of ATP and other nucleotides.

8. Nitrogen from cylinder

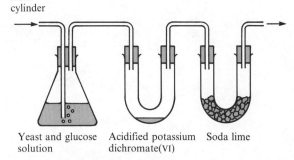

Yeast and glucose solution   Acidified potassium dichromate(VI)   Soda lime

The figure shows an investigation of the respiration of yeast under anaerobic conditions.

(*a*) What will be observed in the acidified dichromate(VI)?

(*b*) What chemical reaction produces this change?

(*c*) The function of the soda lime is to absorb carbon dioxide. Say how the exact amount of carbon dioxide absorbed could be measured.

(*d*) Write an equation for the anaerobic respiration of yeast.

(*e*) Describe how you would modify the apparatus to investigate the aerobic respiration of yeast.

(*f*) What would you expect to observe in the acidified dichromate under aerobic conditions? Explain your answer.

Part 2

FOOD SCIENCE

The subject of **food chemistry** is relatively new. The constituents of food: proteins, carbohydrates, lipids, vitamins and others have been described in Part 1 of this book. They are chemicals, possessing functional groups which will interact with air, water and other food components to produce changes. During the processing of foods, heating, the removal of water and the addition of salts and other substances result in other changes. Chemical reactions can be used to bring about changes which make food more digestible or more appetising. Some chemical reactions, on the other hand, lead to changes which reduce the quality and perhaps the safety of the food. Many people are employed in the food industry as **food scientists** and **food technologists**. Their jobs are to ensure that the changes which the food processing industry makes in foods are for the better.

10

FOOD QUALITY

The food which we eat must provide the energy we need and also the proteins, vitamins and minerals. However, even if a food is nutritious, unless it looks good and tastes good we are not likely to eat much of it. Many of the items in our diet are not nutritious, but we eat them because they have good eating qualities: they have a texture, flavour, odour and colour which please us. We like food to have the characteristics which we expect it to have. We reject food items which are not their natural colour. We expect fruits like tomatoes and strawberries to have a 'bite' and are disappointed in the mushy texture of these fruits when they have been frozen and thawed. We like cooked vegetables to be soft enough to chew easily but firm enough to have a 'bite'.

10.1 WATER CONTENT

Water is bonded by hydrogen bonds to proteins and polysaccharides and by hydration of ions and trapped in proteins and polysaccharides.

A food which is tender and juicy must contain water. Some of the water content is bonded to $>NH$ and $>C=O$ groups of proteins and to $—OH$ groups of polysaccharides by hydrogen bonding. Some water is held by the hydration of ions. A large fraction of the water content is not chemically bound: it is believed to be trapped. Long chains of protein molecules link with one another to form a network, and water is trapped between the chains. Polysaccharide chains also join to form a network of chains which can hold water molecules. The network structure of proteins is affected by extremes of pH and temperature which lead to denaturation, and this is accompanied by loss of water retention. The network of polysaccharide chains can be broken by extremes of pH and by heating.

10.2 TEXTURE OF PLANT FOODS

Plant foods derive their texture from the plant cell wall which is composed of cellulose and pectic substances.

A plant cell is surrounded by a cell surface membrane and a cell wall [see Figure 1.1A]. The cell surface membrane is composed of phospholipids and proteins [§4.12] and is semi-permeable. It is the cell wall which gives a plant cell its firm structure and is responsible for the texture of plant foods. When fruits such as tomatoes and strawberries are frozen and then thawed, they have a soft, mushy texture which is unappetising. The water content has expanded on freezing and broken cell walls, destroying much of the firm structure of the fruit.

The cell wall is composed of carbohydrates, mainly cellulose [see §3.5] and pectic substances, and is permeable to solutes. From pectic substances can be extracted pectin, which is a polymer of α-D-galacturonic acid (related to galactose, with —CO_2H replacing —CH_2OH) and the methyl ester of this acid.

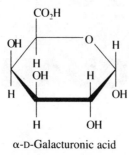
α-D-Galacturonic acid

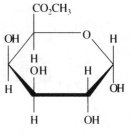
Methyl α-D-galacturonate

The number of units in the pectin chains varies from 50 to 2000. Pectins are added to jams to give them a firmer, less runny texture.

10.3 MEAT

Mammals have three types of muscle:

● Cardiac muscle, which occurs only in the heart

● Smooth muscle, which makes up the walls of structures such as the alimentary canal and blood vessels and is not under voluntary control

● Skeletal muscle, attached to bone, which moves the limbs and body and is under voluntary control, that is, it will contract only when stimulated by a motor nerve.

Most of the animal tissue which we eat is **skeletal muscle**. Muscles are bundles, up to 10 mm in diameter, of **muscle fibres**, each of which is about 0.1 mm in diameter and ranges in length from a few millimetres to several centimetres. Each fibre in skeletal muscle runs the whole length of the muscle and is joined at its ends by tendons to bones or organs [see Figure 10.3A]. A muscle fibre is cylindrical, with a surface membrane (sarcolemma), many nuclei and the regular pattern of bands which is distinctive of muscle fibres. Each fibre is composed of numerous **myofibrils** which lie parallel to one another. Details of muscle structure are shown in Figures 10.3B–E.

Most of the meat we eat is skeletal muscle. Muscles consist of bundles of muscle fibres each of which consists of myofibrils.

FIGURE 10.3A
A Muscle

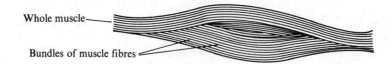
Whole muscle

Bundles of muscle fibres

FIGURE 10.3B
A Muscle Fibre Examined under a High-power Microscope

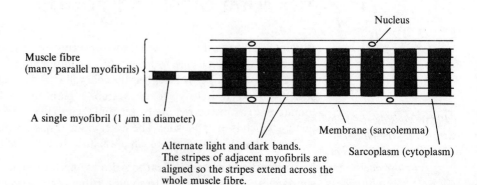

Nucleus

Muscle fibre (many parallel myofibrils)

A single myofibril (1 μm in diameter)

Membrane (sarcolemma)

Sarcoplasm (cytoplasm)

Alternate light and dark bands. The stripes of adjacent myofibrils are aligned so the stripes extend across the whole muscle fibre.

FIGURE 10.3C
A Myofibril under Higher Magnification

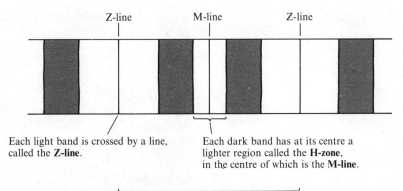

Each light band is crossed by a line, called the **Z-line**.

Each dark band has at its centre a lighter region called the **H-zone**, in the centre of which is the **M-line**.

The repeating unit in the pattern of bands is called a **sarcomere** (2.5–3.0 μm).

FIGURE 10.3D
A Myofibril seen under an Electron Microscope

The myofibrils have a repeating pattern of light and dark bands. Under an electron microscope, the bands are seen to consist of thick myosin filaments and thin actin filaments.

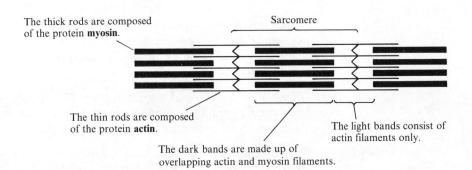

The thick rods are composed of the protein **myosin**.

Sarcomere

The thin rods are composed of the protein **actin**.

The dark bands are made up of overlapping actin and myosin filaments.

The light bands consist of actin filaments only.

Movement takes place when muscle fibres contract or relax. The question of how muscles contract and relax intrigued many people. Two independent research groups, H.E. Huxley with J. Hanson and A.F. Huxley with R. Niedergerke solved the problem. Their microscope studies of muscles in contracted and relaxed states showed that the dark bands were the same length in both. This observation suggested to them that filaments of actin and myosin must in some way slide past one another. In 1954, the two groups independently put forward the **sliding filament theory**. It is illustrated in Figure 10.3E.

The mechanism by which myofibrils contract is a chemical interaction between myosin and actin filaments. When this happens, the filaments slide past one another to overlap more. The actin filaments and therefore also the Z-lines to which they are attached are pulled towards one another, sliding over the myosin filaments. To

FIGURE 10.3E
Muscle Contraction: the Sliding Filament Theory

Muscle contraction takes place as actin filaments slide over myosin filaments, and relaxation is the reverse movement. An animal with contracted muscle makes tough meat.

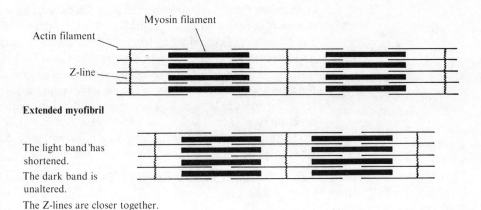

Myosin filament

Actin filament

Z-line

Extended myofibril

The light band has shortened.

The dark band is unaltered.

The Z-lines are closer together.

Contracted myofibril

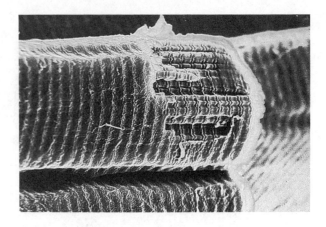

explain how actin filaments are able to slide over myosin filaments, an elegant 'ratchet mechanism' has been proposed. Details are outside the scope of this book but can be found in e.g. *Understanding Biology for Advanced Level* by G. Toole and S. Toole (Stanley Thornes).

Increased contraction makes meat tougher. Muscles contract during exercise and therefore, if an animal is killed in a hunt or if it is in an excited state before slaughter, the muscles have contracted and the meat is tougher.

Muscles are attached to bones by tendons of connective tissue...

Skeletal muscle is attached to bone by means of **tendons**, which are composed of **connective tissue**. Connective tissues are composed of two types of protein fibre: **collagen fibres** and **elastin fibres**. In meat the connective tissue which surrounds the bundles of muscle fibres is mainly collagen. The collagen fibre has the form of an extended spring which cannot be stretched further [see Figure 2.9B]. The elastin fibre has a similar extended spring form and also a coiled spring form, in which the helix turns at shorter intervals. Both collagen and elastin are insoluble and difficult to digest. When meat is cooked (or treated with dilute acid), collagen and elastin are converted into the protein **gelatin**, which is more soluble and readily digestible. Cross-linking between polypeptide chains increases with age and with exercise. This is why meat from older animals is tougher.

...composed of collagen fibres and elastin fibres...

...which on cooking form gelatin.

After the death of an animal, its muscles stiffen and after a day or two they relax.

Some hours after animals die, rigor mortis sets in. The tissues stiffen because the actin and myosin filaments have become cross-linked. After a day or two, the muscles become soft again because actin filaments are detached from the Z-line by enzyme activity. Butchers aim to provide the conditions for this reaction to occur so that the meat will be tender [see § 11.4].

Two types of muscle fibre are fast fibres or twitch fibres – white fibres – which are used for rapid motion and can function anaerobically for a time, and slow fibres or tonic fibres – red fibres – which are used for sustained motion and contain oxymyoglobin, which is red.

There are two types of skeletal muscle fibre: **fast fibres** or **twitch fibres**, which are also called **white fibres**, and **slow fibres** or **tonic fibres**, which are also called **red fibres**. Fast muscle fibres are used for rapid motion. They depend on the anaerobic process of glycolysis for ATP production [see § 9.6] and quickly build up an oxygen debt. Slow muscle fibres are used for steady, sustained motion. They depend on aerobic respiration for their ATP production, although, if oxygen is in short supply, they may function anaerobically to form lactic acid and incur an oxygen debt. Slow muscle fibre therefore needs to have a store of oxygen so that it can function briefly even when deprived of oxygen. The storage substance is myoglobin, which is red when oxygenated (see above). This is why slow muscle fibre is red while fast muscle fibre, which does not need this oxygen store, is pale in colour.

Fish have muscles which are built of fast muscle fibres and are therefore white. Poultry do not exercise their wings and therefore have white breast meat and pinker leg meat. Game birds fly so they have myoglobin in their pectoral muscles and have darker meat. Domestic animals have slow muscle fibre and their meat is red.

10.4 FLAVOUR

There are four primary tastes: **sweet**, **bitter**, **sour** and **salty**.

Sweetness is due to sugars, e.g. glucose, sucrose, fructose. Sugar substitutes, e.g. saccharine, are used as sweeteners.

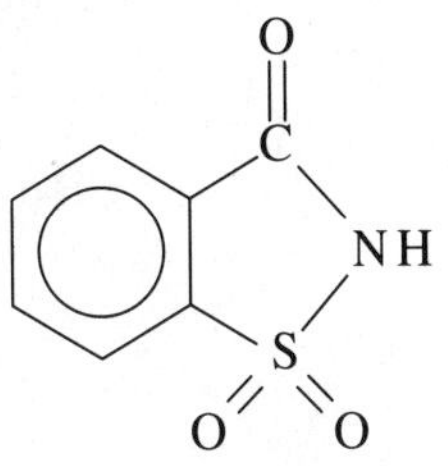

Saccharine

Bitterness is due to a variety of unrelated substances, including caffeine, quinine, calcium ion, magnesium ion and ammonium ion.

A sour taste is due to the presence of an acid, e.g. citric acid, malic acid, lactic acid.

A salty taste is due to the presence of sodium chloride.

Flavour enhancers are substances which, although they have very little taste themselves, intensify the flavours of other food components. Monosodium glutamate is the most widely used [see § 13.2].

Foods may taste sweet, bitter, sour or salty and may have added flavour enhancers.

$$HO_2C \diagdown$$
$$CHCH_2CH_2CO_2{}^- Na^+$$
$$H_2N \diagup$$

Monosodium glutamate

10.5 ODOUR

Odour   Volatile substances are responsible for the odours of foods. The connection between the odour of a compound and the size and shape of its molecules is unknown.

10.6 COLOUR

The main pigments are **porphyrins**, **carotenoids** and **flavonoids**.

The **porphyrins** include the pigments chlorophyll (green) and haemoglobin (red). The structures of chlorophyll and haem are similar, except that chlorophyll contains a magnesium ion and haem contains an iron(II) ion [see Figure 10.6A]. The structure of haemoglobin is shown in Figure 2.9A.

The **carotenoids** are a set of yellow, orange and red pigments.

The **flavonoids** are a set of yellow pigments which occur in many plants and animals.

FIGURE 10.6A
Some of the Pigments in
Food

Chlorophyll

*Pigments in foods include
porphyrins, carotenoids
and flavonoids.*

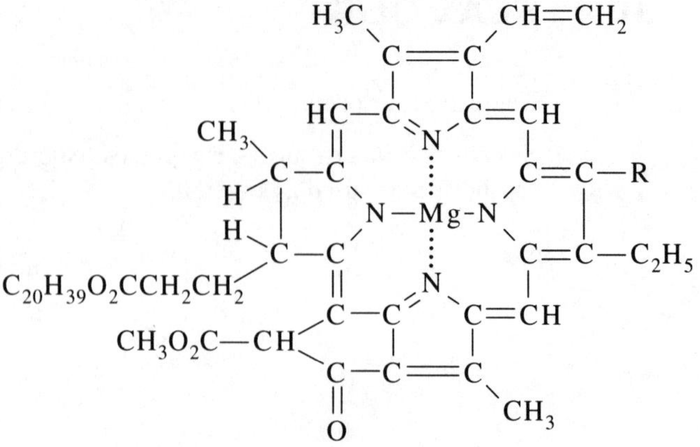

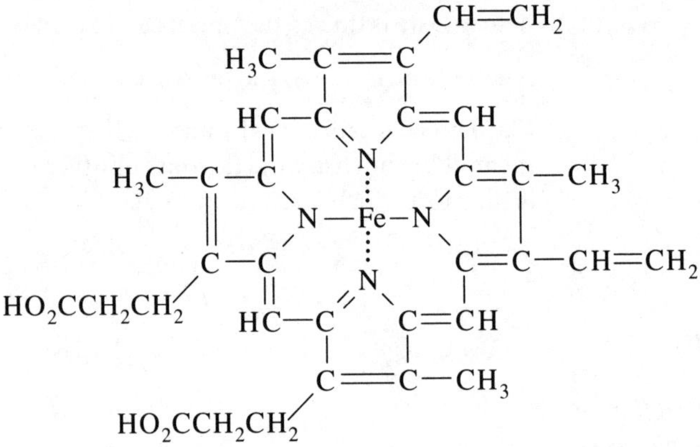

Haem group of haemoglobin
[for haemoglobin see Figure 2.9A]

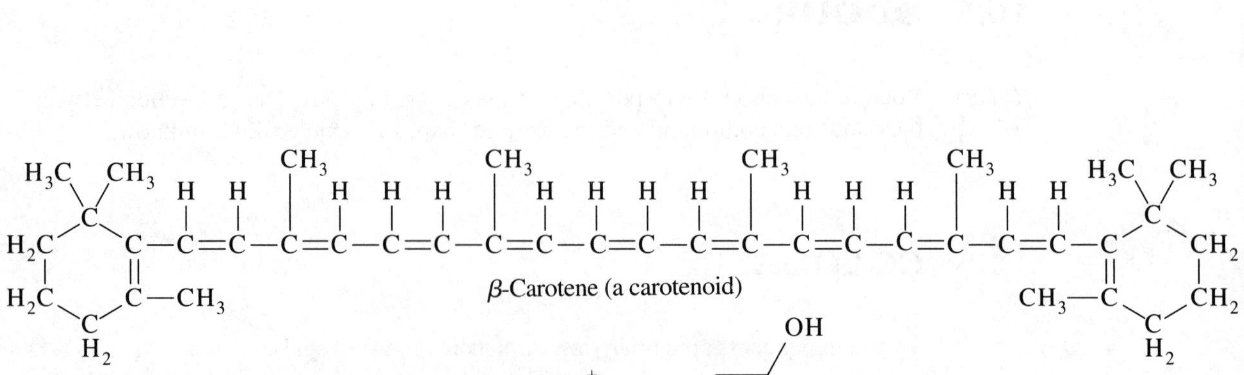

β-Carotene (a carotenoid)

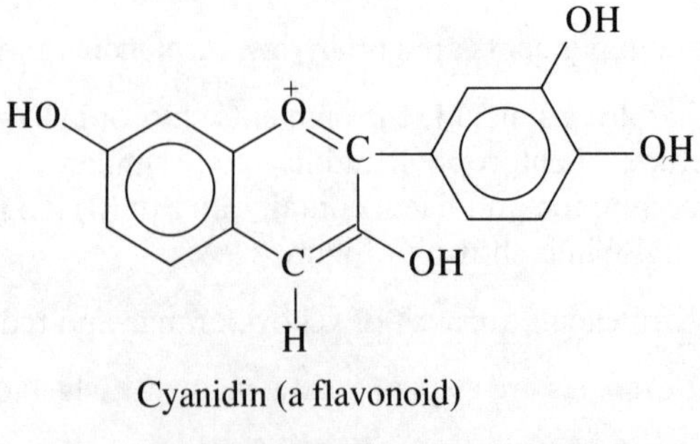

Cyanidin (a flavonoid)

10.6.1 GREEN VEGETABLES

The green colour of vegetables is due to chlorophyll which is hydrolysed by cooking in water.

Although pigments do not have nutritional value, they add to the appeal of foods. The pigments are soluble in water, thermally unstable and affected by extremes of pH. As green vegetables are cooked, the green colour fades. The rate and extent of the colour change depend on the pH of the cooking water. When chlorophyll is heated in acidic solution, the magnesium ion is removed to leave a brownish-green compound. On further heating or in a more strongly acidic solution, the $C_{20}H_{39}$— group (the phytyl group) is hydrolysed off and a brown compound remains. If chlorophyll is heated in alkaline solution, the phytyl group is removed, but the magnesium ion remains, and an olive-green compound is formed. This is the origin of the practice of adding sodium hydrogencarbonate to cooking water. However, the alkali produces other effects also [see § 6.2].

10.6.2 BROWNING OF FRUITS AND VEGETABLES

Many fruits and vegetables, e.g. apples and potatoes, turn brown at the surface after they are peeled. The change is due to the enzyme-catalysed oxidation of compounds in the food. These compounds are polyphenolic compounds which are oxidised to quinones. Further reactions and polymerisation lead to the formation of brown pigments. An example is:

Benzene-1,2-diol (catechol) $+ \frac{1}{2}O_2$ $\xrightarrow{\text{polyphenoloxidase}}$ Cyclohexadiene-1,2-dione (a quinone) $+ H_2O \longrightarrow$ Brown pigments

Browning of fruits and vegetables is due to enzyme-catalysed oxidation of compounds in the food. It can be delayed by reducing agents, e.g. vitamin C.

The browning reaction can be delayed by reducing agents, e.g. sulphur dioxide and ascorbic acid. Ascorbic acid (vitamin C) is a good choice because it is completely harmless and is a nutrient. Browning can be delayed by reducing the activity of the enzyme, polyphenoloxidase. This is done by immersing the food in an acidic solution because enzymes work best over a pH range of 6–8. Sulphur dioxide and sulphites are widely used as preservatives. Weak acids which occur naturally in fruits and vegetables, e.g. citric acid and malic acid, are preferred because they have little effect on flavour. Soaking the food in a concentrated solution of salt or sugar denatures the enzyme and delays browning. Salt is more useful for vegetables and sugar for fruits.

QUESTIONS ON CHAPTER 10

1. (*a*) Mention two types of substance in food which bind water by hydrogen bonding.

(*b*) Draw a structural formula for one of these substances, and show how hydrogen bonding occurs.

2. (*a*) What plant structure is chiefly responsible for the texture of foods of plant origin?

(*b*) What compounds are the chief components of this structure?

(*c*) What animal tissue is chiefly responsible for the texture of meat?

3. Define or explain the terms (*a*) skeletal muscle, (*b*) myofibril, (*c*) sarcomere, (*d*) Z-line, (*e*) H-zone, (*f*) M-line

4. Give a brief description of the sliding filament theory of muscle contraction.

5. (*a*) Explain the differences between fast and slow muscle in
(i) the function they perform and
(ii) their metabolism.

(*b*) Explain why beef is red meat while chicken breast is white meat.

11

FOOD PROCESSING

When foods are heated, the many different chemical reactions, including enzyme-catalysed reactions which are occurring in the foods, take place at an increased rate. If the temperature is raised above about 60 °C, enzymes are denatured and lose their catalytic activity. Some enzymes are more temperature-sensitive than others and lose their activity above about 40 °C.

11.1 COOKING VEGETABLES

In § 10.2, we looked at the structure of plant cell walls. Important components in maintaining the structure of vegetables are the pectin polysaccharides in the cell walls. When vegetables are heated in water, some of the —OCH_3 groups in the pectin molecules are hydrolysed and the pectins become more soluble. When the pectins become more soluble, the texture of the vegetable becomes softer. This is the desired effect of cooking vegetables, but over-cooking makes the vegetables unpleasantly soft and mushy. When hard water is used for cooking vegetables, the pectins from which methoxy groups have been hydrolysed can react with calcium and magnesium ions in the hard water. These ions can link pectin molecules together through their carboxyl groups:

When vegetables are heated in water, pectins are partially hydrolysed to become more soluble, rendering the vegetables softer.

$$—CO_2^-\ \ Ca^{2+}\ \ ^-O_2C—$$

In this way, the texture remains firmer than that of vegetables cooked in soft water. Alkaline cooking water both increases the hydrolysis of pectins and also removes calcium ions as sparingly soluble calcium compounds and both these effects make vegetables less firm. The practice of adding sodium hydrogencarbonate to green vegetables to preserve their colour therefore softens the vegetables and also decreases their nutritive calcium content by removing calcium as calcium hydrogencarbonate and calcium carbonate.

Cooking in hard water retains a firmer texture. Alkaline cooking water removes nutrients.

A loss of vitamins occurs on boiling vegetables, especially thiamin (vitamin B_1) and ascorbic acid (vitamin C) which have a high solubility in water and are unstable at cooking temperatures.

Vitamins are lost in cooking.

The main pigments responsible for the colours of foods are porphyrins, carotenoids and flavonoids. The porphyrins include chlorophyll [see Figure 10.6A]. Carotenoids are yellow, orange or red fat-soluble pigments, and flavonoids are yellow water-soluble pigments [see Figure 10.6A].

The compounds which colour plants and animals are water-soluble, decompose when heated and are affected by high and low values of pH. When vegetables are cooked,

Pigments in plants and animals are water-soluble and decompose when heated.

the brightness of the green colour fades, changing to olive green, then yellow-green and finally to a brownish green. The rate and extent of the colour change depend on the pH of the cooking water [see § 10.6]. To preserve the green colour of vegetables, cooks sometimes add sodium hydrogencarbonate to the cooking water, but this is not a wise move because it decreases the calcium content as described above.

Microwave cooking retains nutrients and colour.

With the increasing popularity of microwave ovens, more people are microwaving their vegetables. With this method of cooking, the problem of vitamins and minerals dissolving in the cooking water does not arise. In addition, there is no loss of colour.

CHECKPOINT 11.1

1. Refer to the structure of pectin [see § 10.2]. (*a*) Show by means of a diagram how a calcium ion can form a cross-link between two molecules of pectin.

(*b*) Explain why sodium ions cannot achieve the cross-linking.

2. (a) Describe and explain the effect of heat and pH on the green colour of vegetables.

(*b*) What effect does adding sodium hydrogencarbonate in the cooking of green vegetables have on (i) their appearance, (ii) their nutritional quality?

(*c*) How does the microwave cooking of green vegetables compare with boiling in water
(i) with regard to taste, (ii) with regard to nutritional value?

11.2 MAKING FLOUR

FIGURE 11.2A
A Wheat Grain

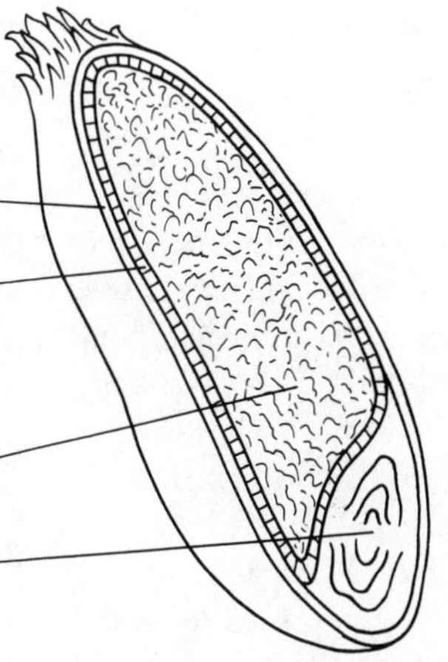

Tough outer layer (bran) consists of two layers:

Husk or fruit coat (pericarp) fused with seed coat (testa) on the outside is mainly cellulose.

Aleurone on the inside contains fat and protein.

Endosperm is mainly starch, with some protein.

Embryo

The wheat grain is the fruit of the plant and is a food source. It contains starch, protein, cellulose, some fat and small amounts of vitamins and minerals. Wheats are described as **hard** or **soft** according to the texture of the endosperm. Hard wheats contain more protein than soft wheats. Wheat grown in the UK has a relatively low protein content (8%) and is blended with other wheats, e.g. North American wheat, to give a protein content of 10–15%. A hard wheat gives a strong flour, while a soft wheat gives a weak flour [see § 10.2.2]. In making white flour, the husk (fruit wall) and the germ (the embryo plant) of the grain are removed. In wholemeal flour, the whole of the wheat grain is ground into flour.

Wheat grain is the fruit of the plant and contains protein, cellulose, fat and small amounts of vitamins and minerals.

11.2.1 MILLING

Flour is made by grinding wheat grains in a mill. Before wheat is milled, it is conditioned to bring the water content to 15%. After conditioning, different wheats

Flour is made by milling wheat. The grains separate into bran, endosperm and flour. The endosperm breaks into fine particles.

are gristed (blended together) so that they will give flour of the required protein content. In milling, the wheat grains are passed through **break rollers** which roll open the grains and break them up into bran, bran-free chunks of endosperm and flour. The bran is removed and the pieces of endosperm are passed through **reduction rollers** which break the endosperm into fine particles. The germ and bran particles are removed by sieving.

The whitest flour separates in the early stages.

The process is repeated up to 12 times. After each reduction roll, the flour is separated from the bran by sieving, and the particles which remain go on to the next roller. It is possible to extract about 65% of the grain as white flour. The whitest flour is produced in the early stages of milling. If the mill is set to scrape more flour from the bran, some bran particles will escape into the flour and the flour will not be as white as required. The miller collects different streams of flour and blends them to give different varieties of flour.

11.2.2 DOUGH-MAKING

When flour is mixed with water to make dough, the proteins form a complex mass called **gluten**. If the dough is held in a stream of water, the starch is washed away to leave gluten. Gluten proteins consist of **gliadins** and **glutenins**. The gliadin proteins have relative molecular masses of 30–40 000 and have low solubility. The glutenins have relative molecular masses of 40 000–20 million and very low solubility. Glutenins are able to form an extended three-dimensional network of coiled protein chains linked by disulphide bridges. A high proportion of glutenins in the flour give a dough which is 'strong', needs more mixing and gives loaves of greater volume.

In dough, the proteins, gliadins and glutenins form gluten, a three-dimensional network of protein chains.

The protein content of grain can be measured by the Kjehldahl method [see § 2.5]. An approximate method of assessing protein is carried out at mills. The grain is ground milled and sieved. A little of the flour is mixed with water to make a stiff dough. The dough is left in a stream of cold running water until all the starch has been washed away as a cloudy suspension. The water-insoluble gluten remains. The size of the gluten ball is a rough measure of protein content, and its colour, elasticity and toughness can be assessed.

To assess protein, wash away the starch from a ball of dough and observe the size of the gluten ball that remains.

Gluten provides the solid structural framework which can trap the gases in bread. A hard wheat gives a **strong flour** with a high level of protein. This gives a good yield of gluten, which is strong enough to trap the large volume of gas required to leaven bread.

A hard wheat gives a strong flour, used for bread. A soft wheat gives a weak flour, used for cakes etc.

A soft wheat gives a **weak flour** with a low protein level, which gives a more elastic gluten, which is needed for biscuit, cake and pastry production.

The bread-making properties of a flour improve with prolonged storage. During storage, oxidation of polyunsaturated fatty acids in flour results in the formation of peroxides. These oxidising agents bleach the pigments in the flour, giving the bread a whiter crumb. Prolonged storage also results in the oxidation of some of the free —SH groups of the gliadins to —SO_3H, which increase the elasticity of the dough. The baking properties of the flour are thus improved by ageing. Instead of storing flour for a long time to autoxidise, manufacturers prefer to bleach it with chlorine or chlorine dioxide. Chemical ageing destroys some of the nutrients in flour.

The bread-making properties of flour improve with ageing, natural or chemical.

A good dough can retain gas when it is baked. Glutenin forms a three-dimensional structure that will hold gas.

A good dough is one that will hold a large volume of gas and retain it when the protein content of the dough sets during baking. A dough is made by mixing and kneading the ingredients. During mixing, the glutenin molecules become stretched into chains which interact to form sheets round the gas bubbles. The stretching of glutenin involves chemical reactions. Hydrogen bonds are broken and reformed as disulphide bridges are broken and remade. Flour improvers such as potassium

Flour improvers oxidise —SH groups, reduce the number of disulphide bridges and weaken the glutenin structure so it can expand more and hold more gas.

bromate(V) and potassium peroxodisulphate(VI), $K_2S_2O_8$, oxidise some of the —SH groups to —SO_3H, reducing the number of disulphide bridges and enabling the dough to expand more readily and give a greater volume of loaf. Ascorbic acid is a flour improver. An enzyme in flour, ascorbic acid oxidase, catalyses the oxidation of ascorbic acid to dehydroascorbic acid, which is an oxidising agent that can oxidise —SH to —SO_3H.

Wholemeal bread contains bran and wheatgerm. White flour has added vitamins and minerals.

In making white flour, the bran and germ are removed from the wheat grain, and with them vitamins are lost. Wholemeal flour contains the bran, which has a high cellulose content and therefore provides roughage to assist the passage of food through the intestines. The wheat germ contains oil and therefore wholemeal flour does not have as long a shelf-life as white flour. White flour is enriched with vitamins B_1, B_2 and B_3, iron and calcium (as calcium carbonate). [See § 7.2 for the calcium content of bread.]

11.2.3 TESTING FOR FLOUR IMPROVERS

Testing for flour improvers:

In the following tests for common flour improvers, control experiments should be done.

BROMATE(V)

1. Press a sample of flour into a Petri dish. Immerse the dish in water for a minute.

...bromate(V)...

2. Pour a solution containing potassium iodide and bench sulphuric acid over the wet flour.

3. Purple or black spots of iodine appear if a bromate(V) is present. Iodate(V) is not permitted as a flour additive in the UK.

ASCORBIC ACID

1. Press a sample of flour into a Petri dish. Immerse the dish in water for a minute.

...ascorbic acid...

2. Pour a solution of iodine in potassium iodide solution over the wet flour.

3. The flour turns blue as starch reacts with iodine. Ascorbic acid is a reducing agent. By reducing iodine it produces white flecks on the blue surface of the flour.

IRON

1. Press a sample of flour into a Petri dish.

...iron.

2. Prepare a solution of potassium thiocyanate and bench hydrochloric acid. Add it to the flour, and leave to stand for 20 minutes.

3. Flour which contains iron shows deep red spots.

11.3 BREAD MAKING

11.3.1 GELATINISATION

When starch is heated with water, gelatinisation occurs. Starch granules take up water, swell and burst to release a suspension of starch.

Wheat flour contains amylases which can catalyse the hydrolysis of amylose and amylopectin in starch. The hydrolysis occurs to only a very slight extent in dry flour, but it begins immediately a dough is made with the formation of maltose, a disaccharide. When a dough is heated, starch granules take up water. When the temperature rises to the **gelatinisation temperature**, the granules suddenly swell and take up a large amount of water. Soluble starch molecules begin to leak out of the granules, and at a higher temperature, the granules rupture, releasing more free starch. The process is called **gelatinisation**. This gelatinised starch is more readily hydrolysed by the starch-digesting enzymes in the human digestive system. The gelatinisation temperature of a starch can be found as shown in Figure 11.3A.

FIGURE 11.3A
Finding the Gelatinisation
Temperature of Starch

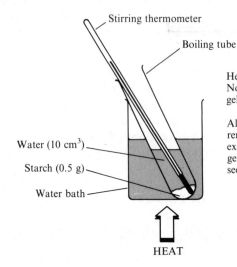

Stirring thermometer

Boiling tube

Heat and stir until the suspension thickens. Note the temperature; this is the gelatinisation temperature of the starch.

Alternatively, at intervals of 5 °C, remove a drop of the suspension and examine it under a microscope. At the gelatinisation temperature, you will see the starch grains swell and burst.

Water (10 cm^3)

Starch (0.5 g)

Water bath

HEAT

11.3.2 LEAVENING

Leavening is the action of gases on bread dough. The action of yeast on glucose produces carbon dioxide, ethanol vapour and water vapour.

When a dough of flour and water is baked, unleavened bread is obtained. Modern bread is lighter than this because it contains many pockets of carbon dioxide. **Leavening** is the action of gases on bread dough, making it rise. Leavening agents are yeast, chemical rising agents, air and steam. Enzymes in yeast convert the sugars naturally present in the flour and the maltose, which has been produced through the action of amylases, into glucose. Glucose is fermented by yeast [§ 9.6] producing carbon dioxide, which makes the dough rise, and ethanol, which vaporises and assists in the leavening.

Sodium hydrogencarbonate is a chemical rising agent.

The most important chemical rising agent is sodium hydrogencarbonate (baking soda) which releases carbon dioxide both on heating and on reaction with an acid. Baking powder contains sodium hydrogencarbonate and a mixture of weak acids, e.g. tartaric acid and sodium aluminium sulphate, which is a slow-acting acid that promotes slow and continuous leavening.

11.3.1 BAKING

Mass production in a bakery involves: mixing the ingredients, fermentation, kneading the dough, fermentation, division into loaves, rising, baking, death of yeast, gelatinisation of starch, coagulation of gluten to form a solid loaf containing leavening gases.

A bakery makes dough in a large mechanical mixer from flour, water, yeast, salt, sugar and fat. The dough is left to ferment for an hour at 25 °C. It is kneaded to expel some carbon dioxide and to bring the yeast into contact with more sugars. After further fermentation, the dough is divided into loaves and allowed to rise again (to **prove**). It is then baked at 120 °C for 30–50 minutes. The dough at first expands rapidly as the carbon dioxide and air in the dough expand. Ethanol vaporises and assists in the leavening. At first the yeast becomes more active as the temperature rises, and more gas is produced. At about 55 °C, the yeast is killed and fermentation stops. Water causes starch grains to swell and gelatinise. The structure of the loaf is supported by gelatinised starch. At about 75 °C, gluten begins to coagulate [see § 11.2] to form a solid structure which contains the leavening gases, and this process continues to the end of the baking period. Water vapour, ethanol vapour and some of the carbon dioxide escape during baking. The reason why salt is added to the dough is that it strengthens the gluten, possibly by inhibiting proteolytic enzymes which could attack gluten.

FIGURE 11.3B
(a) A Mixer, (b) a Prover,
in a Bakery

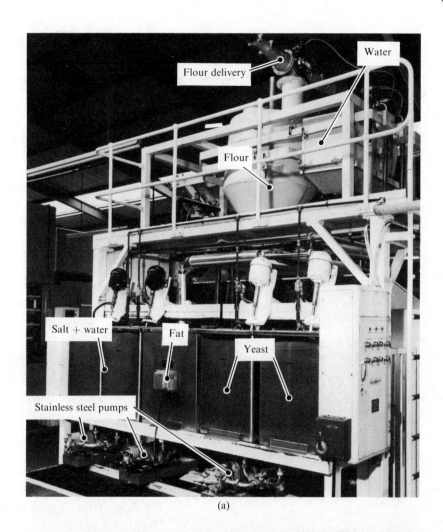

(a)

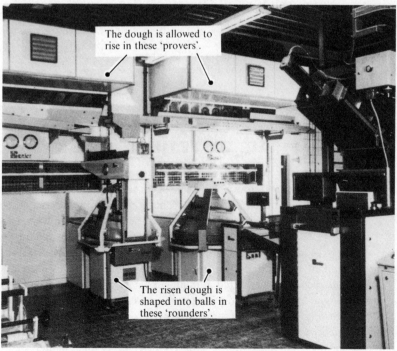

(b)

11.3.4 BROWNING

Browning is due to caramelisation of sugar ...

The high temperature at the surface of the bread causes browning to occur. It is due to two non-enzymic processes. One is the caramelisation of sugar:

$$C_6H_{12}O_6(s) \rightarrow 6C(s) + 6H_2O(l)$$

The other is a reaction between a reducing sugar and an amine. This produces more rapid and more intense browning and a savoury smell. It takes place when a reducing sugar is heated with an amine such as an amino acid or a protein containing lysine residues. This type of browning is called a **Maillard reaction**. It occurs at the surface of dry-cooked foods containing protein and sugar, e.g. bread, custard, milk pudding and roast meat. There may be some loss of nutrient: up to 15% of the lysine in bread can be lost by crust-browning.

... or a Maillard reaction between a reducing sugar and an amine.

Caramel is used for food colouring. It used to be produced by heating sugar alone, but it is now made by heating sugar in the presence of ammonia, that is, by the Maillard reaction.

CHECKPOINT 11.3

1. What is each of the following? (*a*) endosperm, (*b*) gluten, (*c*) leavening, (*d*) Maillard browning.

2. (*a*) What grain is chiefly used for flour making?

(*b*) What is the difference between a hard wheat and a soft wheat?

(*c*) What is the difference between a strong flour and a weak flour?

(*d*) What function does gluten perform in bread-making?

(*e*) Why is sugar added to bread dough?

(*f*) Name three leavening agents.

3. A lump of bread dough contains $20 \, cm^3$ of air at room temperature. What will be the volume of air at $100 \, °C$? (For help, see *ALC*, §7.2.3.)

4. A lump of dough contains $18 \, cm^3$ of water at $20 \, °C$. What volume of water vapour will be present when the dough is heated to $100 \, °C$?

5. Eclair shells are highly leavened. Ammonium hydrogen-carbonate is added to the baking powder, and a very hot oven is used. Explain why these two factors assist in leavening.

6. Three sacks have been delivered to the bakery. The labels have come off. One label reads 'Hard wheat flour', one reads 'Soft wheat flour', and the third reads 'Self-raising flour'. Describe how you would test to find out which flour is which.

7. (*a*) What is the gelatinisation of starch?

(*b*) Describe how the gelatinisation temperature can be found.

(*c*) What importance does gelatinisation have in digestion?

(*d*) How can a specimen of gluten be obtained from flour?

(*e*) What importance does gluten have in the baking of bread?

11.4 THE BUTCHER'S TASK

After the death of an animal, anaerobic respiration of glucose continues, with the formation of lactic acid which slows the spoilage of meat.

After an animal has been slaughtered, its oxygen supply is cut off and the tissues become anaerobic. The enzymes continue to metabolise glucose anaerobically to lactic acid [see §9.3]. Since lactic acid, like other acids, slows down the growth of micro-organisms, it slows spoilage and increases the time for which meat will keep. If an animal is slaughtered in a well-fed, rested condition, its tissues contain more glycogen and glucose than one that is hungry or has been killed after a chase. Meat from a slaughtered animal will have a higher lactic acid content and keep better than meat from a hunted animal. After death, rigor mortis sets in, and meat becomes quite tough. If meat is allowed to hang, proteolytic enzymes tenderise the meat [see §10.3]. Animals may be injected with proteolytic enzymes before slaughter so that enzymes reach all the tissues. One can buy meat tenderisers to use in the home. They contain the proteolytic enzyme papain. The meat is marinaded (soaked) in a solution of the enzyme. When cooking starts, the rising temperature increases the activity of papain and the meat is tenderised. As the temperature rises higher, the enzyme is denatured.

Meat is hung to allow proteolytic enzymes to tenderise it.

Commercial tenderisers contain papain.

When meat is handled, it is important to preserve the red colour which the consumer takes to indicate freshness. Meat tends to darken as water is lost from a cut surface, concentrating the pigments and making them look darker. The protein myoglobin is purple because it contains iron(II) ions and haem [see Figure 10.6A]. When a cut surface of beef is exposed to the air, myoglobin is converted into oxymyoglobin, which is bright red. It is necessary to keep myoglobin in the form of oxymyoglobin to maintain the bright red colour which makes the meat attractive to the consumer. If the supply of oxygen is inadequate, the pigment is oxidised to metmyoglobin, which is brown.

$$Mb^+(Fe^{3+}) \quad \xleftarrow{\text{oxidation}} \quad Mb(Fe^{2+}) \quad \xrightarrow{\text{oxygen}} \quad MbO_2(Fe^{2+})$$

Metmyoglobin (brown)   Myoglobin (purple)   Oxymyoglobin (bright red)

Meat contains myoglobin, which is purple. Oxymyoglobin is bright red. To allow oxygen to reach the meat and keep the protein an attractive red colour, meat is packaged in 'breathing film'.

It is a rather strange problem in that it is necessary to maintain an adequate supply of oxygen to keep myoglobin in the form of oxymyoglobin and *prevent* its oxidation to metmyoglobin. The problem arises when meat is cut and packaged: once it is sealed up, aerobic bacteria present in the meat consume some of the oxygen and cause the surface oxymyoglobin to become deoxygenated and therefore susceptible to oxidation to brown metmyoglobin. The solution to the problem is to use packaging which allows oxygen to enter, a type of packaging known as 'breathing film'. Cooking turns myoglobin brown because it oxidises the iron(II) ion at the centre of the molecule.

Sodium nitrite is used to cure meat, converting myoglobin into nitrosomyoglobin, which is pink.

When meat is cured, sodium nitrite is a chief ingredient. It is converted into nitrous acid, which reacts with myoglobin to form nitrosomyoglobin, which gives cured meat a pink colour. Another ingredient in the curing solution is sodium chloride, which inhibits the growth of certain micro-organisms [see § 12.4.6]. Lactic acid bacteria are more tolerant of salt than other bacteria so the formation of lactic acid, which acts as a preservative, is favoured.

QUESTIONS ON CHAPTER 11

1. Define or explain the terms: (*a*) endosperm, (*b*) gluten, (*c*) Maillard browning, (*d*) leavening.

2. (*a*) Why is sugar added to bread dough?

(*b*) Why does bread turn brown on the outside?

(*c*) Name three common leavening agents used in baking.

3. Flour was in the past stored for several weeks before being used to make bread. During the ageing process, the bread-making characteristics of the flour were improved. To save time, bakers now add flour-improvers as a substitute for the ageing process.

(*a*) Explain two functions of flour-improvers.

(*b*) Name two chemicals which are used as flour-improvers. Describe how you could test a sample of flour to find out whether they were present.

4. (*a*) What function does gluten perform in baking?

(*b*) Describe how you could make a rough assessment of the quantity and quality of gluten in a flour.

5. (*a*) What are the differences between hard and soft wheats?

(*b*) What is the gelatinisation temperature of a flour? Describe how you could determine it experimentally.

6. (*a*) Why does meat from a well-fed, rested animal keep better than meat from a hungry, exercised animal?

(*b*) What is the crucial ingredient in a meat marinade?

(*c*) What is the cause of (i) the bright red colour of fresh meat, (ii) the pink colour of cured meats, e.g. ham?

12

FOOD PRESERVATION

12.1 FOOD SPOILAGE

Foods contain enzymes and are therefore susceptible to spoilage.

All food was once part of a living organism. Some foods, e.g. fish and meat, are from organisms which were killed before the food became available. Other foods, e.g. fruits and vegetables, may be gathered and distributed while still in the living state. Food contains enzymes and is therefore susceptible to change and spoilage. About one-fifth of the world's food is lost through spoilage. There are two types of food spoilage: **autolysis** and **microbial spoilage**.

12.2 AUTOLYSIS

Autolysis is the breakdown of foods caused by enzymes within the food. It begins after the animal is killed or the crop is harvested. Some enzyme activity improves food, e.g. meat becomes more tender and fruits ripen. Most autolytic changes, however, cause food to deteriorate. The oxidation of fats gives rancid flavours and smells, which make food unacceptable. Water can be lost from the surface, leaving an unappetisingly dry food.

Autolysis is the deterioration of food caused by enzymes in the food.

When some fruits and vegetables are peeled the fresh surface turns brown fairly rapidly. Enzyme-catalysed oxidation of the exposed tissues takes place. It can be delayed by any of the methods that denature enzymes, e.g. immersion in acidic solution or a concentrated salt solution, blanching (soaking in boiling water). Non-enzymic browning reactions of the Maillard type (see §11.3) also occur. These can be delayed by sulphites.

12.3 SPOILAGE BY MICRO-ORGANISMS

Nutrients released from cells by autolysis are available to micro-organisms.

Autolysis causes the release of nutrients from the cells. Once outside the cells, the nutrients are available to micro-organisms. Bacteria, moulds and yeasts abound in the atmosphere. In contact with autolysing food, they feed and multiply. Eventually the food starts to look 'off'.

MOULDS

Moulds are a type of fungus. They need oxygen and therefore grow on the surface of food as fine threads, hyphae, which form a branched network called a mycelium. A sort of fluff can be seen on the surface of the food. Bread is soon attacked by moulds because its porous structure allows air to enter it. Moulds do not grow in alkaline or strongly acidic solutions; you will not find a jar of pickles going mouldy. Moulds can grow slowly inside a home refrigerator, they grow best at pH 4–6 and a temperature of about 30 °C. Moulds produce spores which can be carried by the wind. A **spore** is a dormant state of a micro-organism which forms as part of the life cycle in asexual reproduction. The life processes have been shut down to the minimum needed to keep the organism alive. Spores can be carried long distances by air currents and in this way infect food. To ensure destruction of all moulds and their spores, food must be sterilised under pressure, i.e. above 100 °C.

Moulds need oxygen and grow on the surface of food. They produce spores.

YEASTS

Yeasts are, like moulds, a type of fungus. They reproduce by budding: a small offshoot or bud grows to a certain size and then separates from the parent yeast cell. Yeasts form spores which are less temperature-resistant than those of moulds and bacteria. Yeasts occur in the soil and on the surface of fruits. The 'bloom' which you can see on the surface of grapes is a yeast which catalyses the fermentation of grapes in wine-making. Yeasts can grow at fairly low pH, in high concentrations of salt and sugar and in the absence of oxygen. Yeasts are used as flavourings in meat pies, sausages, potato crisps and other items. Marmite contains extracts of yeast and is rich in vitamins of the B group.

Yeasts occur in the soil and on the surface of fruits. They produce spores.

BACTERIA

Bacteria are one of the smallest forms of life. One million bacteria weigh less than one millionth of a gram. They include cocci (spherical in shape), bacilli (rod-shaped) and spirilli (spiral in shape). A bacterium absorbs nutrients from its environment, grows until it reaches a certain size and then divides into two new cells. When the number of bacteria reaches ten million per gram of food, the food looks obviously spoiled. Most bacteria grow best in neutral conditions. Some bacteria tolerate fairly low pH, e.g. *Lactobacillus*, which causes the souring of milk with the formation of lactic acid, and *Acetobacter*, which converts ethanol into ethanoic acid. Some bacteria, **aerobic bacteria**, grow only in the presence of oxygen. Other bacteria, **anaerobic bacteria**, grow only in the absence of oxygen. Some bacteria can grow under both aerobic and anaerobic conditions. Bacteria are killed at 100 °C, but some produce spores which are heat-resistant. Some bacterial spores can survive boiling for several hours.

Aerobic bacteria grow in the presence of oxygen, and anaerobic bacteria grow in the absence of oxygen. Some produce spores.

12.3.1 FOOD POISONING

Food-poisoning is due to the release of toxic metabolic products by micro-organisms into food.

As micro-organisms grow on food, they release into the food enzymes which catalyse the breakdown of carbohydrates, lipids and proteins into nutrients which the micro-organisms can use. Some micro-organisms have metabolic products which are toxic to animals that eat the contaminated food. The bacterium *Salmonella* can be present in food without making the food look obviously bad. It produces a toxin *after* the infected food has been eaten. *Salmonella* does not form spores and can be destroyed by heating at 60 °C for 15–20 minutes. *Salmonella* cannot multiply in a closed container (a can or bottle) and it cannot multiply in cold food. When cooked food is kept at room temperature, the conditions are ideal for the growth of *Salmonella*. *Salmonella* is one of the typhoid bacteria. Some people are carriers of typhus: although infected they show no symptoms. If these carriers handle food which is kept under unhygienic conditions the food becomes infected and spreads the disease.

Salmonella can be killed by heating at 60 °C for 15–20 minutes and cannot multiply in a closed container or in cold food.

Staphylococcus is a bacterium which causes pimples, boils and septic wounds. It also causes food poisoning, so people who have septic cuts on their hands should not handle food. A bandage or plaster over a septic cut is not protection because once it becomes wet it does not stop bacteria from reaching the food. *Staphylococcus* produces a toxin in infected food. It does not form spores, and it is quickly killed by boiling. The toxin, however, is more persistent, and food must be kept at 100 °C for 30 minutes to destroy all the toxin. Many foods cannot be heated to such a high temperature, e.g. cream-filled pastries, and these may be a source of food-poisoning by *Staphylococcus*.

Staphylococcus is quickly killed by boiling.

The most dangerous type of food-poisoning is caused by *Clostridium botulinum*. This micro-organism is present in soil and may contaminate vegetables. Someone who has been working in the fields or handling manure can spread the bacterium to food. The toxin produced by the bacterium is called **botulin**. It causes the food poisoning called **botulism**, which may cause paralysis and even death. Botulin is one of the most toxic of substances: an estimate is that half a kilogram would kill the whole of the world's population. The toxin is destroyed by heating, but the micro-organism itself can withstand high temperature. Spores of *Clostridium* are destroyed by heating food at 121 °C for 10 minutes. Food which has not been properly bottled or canned may contain *Clostridium*, which will live in the canned food, producing botulin, which will cause food poisoning when the contents of the can are eaten. Commercial canning and bottling plants employ temperatures which kill *Clostridium*.

Clostridium botulinum is present in soil. It produces the toxin botulin which causes botulism, a form of food poisoning which may be fatal. Its spores are killed by heating at 121 °C for 10 minutes.

Some micro-organisms are beneficial in foods. Microbial growth is necessary for the ripening of many cheeses, e.g. Stilton, Camembert, Rocquefort, and for the maturing of yoghurt.

12.3.2 FACTORS WHICH AFFECT THE GROWTH OF MICRO-ORGANISMS

Figure 12.3A shows the way in which the population of a micro-organism changes with time. When food is contaminated with a micro-organism, there is a lag phase, while the organism adapts to its new environment. There follows a period of rapid growth. In the case of bacteria, the population doubles every 20 minutes. The rate of growth then declines and the population remains constant as nutrients are used up or the organism is inhibited by a build-up of its metabolic products. The food begins to appear unpalatable before the stationary phase is reached.

FIGURE 12.3A
Growth of a Micro-Organism

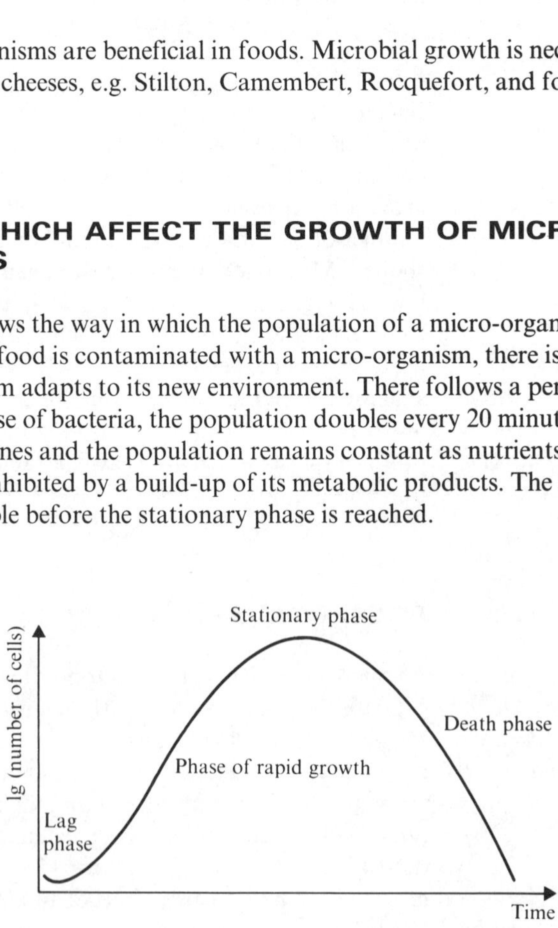

The population of micro-organisms changes with time through a lag phase, a phase of rapid growth, a stationary phase and a death phase.

Figures 12.3B, C and D show the effects of temperature, pH and relative humidity on the growth rates of moulds, yeasts and bacteria.

FIGURE 12.3B
The Effect of Temperature
on Growth Rate

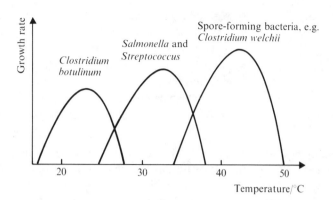

FIGURE 12.3C
The Effect of pH on
Growth Rate

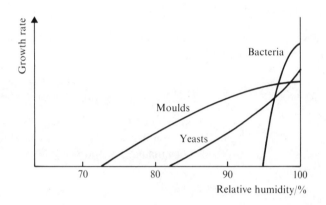

FIGURE 12.3D
The Effect of Relative
Humidity on Growth
Rate

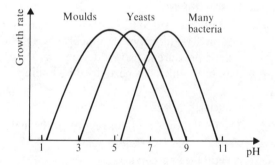

12.4 FOOD PRESERVATION

The conditions which micro-organisms require for growth are:

● Nutrients – to supply energy, to supply nitrogen for protein synthesis and minerals
● Moisture
● A suitable temperature. Some micro-organisms can grow at 50–60 °C; others have optimum growth temperatures of 25–40 °C, and others grow well at 10–20 °C, and can survive at 0 °C.
● Time. Some bacteria divide once every 20 minutes. If the food supply were adequate, a single bacterium could produce 2 million offspring in 7 hours. In practice, however, the growth rate declines as the food supply is exhausted.

● A suitable atmosphere. Some organisms, e.g. moulds and some bacteria, are aerobic. Yeasts and many bacteria can thrive in either the presence or the absence of oxygen. Some organisms, chiefly bacteria, are anaerobic.

● pH. Most micro-organisms grow best at a pH of 6.6–7.5. Some micro-organisms grow at lower pH values. The spoilage of fruits is usually caused by yeasts and moulds, which can tolerate low pH. The spoilage of meat and fish is usually caused by bacteria.

Food preservation depends on interfering with the conditions which micro-organisms need for growth.

In order to keep food wholesome, the growth of micro-organisms must be prevented. Some preserving methods are **bactericidal**: they kill all the micro-organisms. They do this either by eliminating one of the conditions needed for growth or by using chemicals to kill the micro-organisms. Other methods are **bacteriostatic**: they reduce the activity of micro-organisms to a low level without actually killing them.

12.4.1 REMOVAL OF WATER

The **water activity** of a food describes the amount of available water in a food. It is given by:

$$\text{Water activity} = \frac{\text{Water vapour pressure above the food at a certain temperature}}{\text{Water vapour pressure above pure water at the same temperature}}$$

The water activity of a food must be reduced to 0.6 to prevent the growth of micro-organisms. A number of techniques are used for reducing the water content of food to this level. They include the following.

ROLLER DRYING

The food, e.g. potato flakes, is passed over heated rollers.

SPRAY DRYING

A solution or suspension, e.g. milk, is sprayed into a heated container at reduced pressure. Water vaporises quickly since the air pressure around it is below atmospheric pressure to leave a dry powder. Milk powder made in this way tastes better than milk which has been dried by boiling.

TUNNEL DRYING

The food moves on a conveyor belt through a warm tunnel.

FLUIDISED BED DRYING

The food, e.g. vegetables, is chopped up and put into a column. Warm air blowing through the column from the bottom keeps the particles of food in motion.

FREEZE DRYING

The food is frozen quickly and then warmed slightly under reduced pressure so that the ice crystals sublime. Freeze-dried foods, e.g. instant coffee, contain practically no moisture and can be stored for a long time at room temperature. The method of drying leaves a porous structure, which makes them easier to rehydrate than heat-dried foods. Freeze-dried foods are, however, fragile and need careful handling. The open structure exposes the food to oxidation, and if fats are present they may deteriorate rapidly. To avoid this, freeze-dried foods are often packed under nitrogen. This method of storage enables foods to keep for two or three years.

Before vegetables are dehydrated or stored in any other way they are blanched (scalded) in boiling water or steam. **Blanching** inactivates enzymes such as catalase and ascorbic acid oxidase, thus improving the appearance of the preserved vegetables and increasing their storage life.

SMOKING

Warm air containing smoke is passed over the food. This is one of the oldest methods of preservation. Smoking changes the taste of the food, e.g. kippers, smoked salmon.

OSMOTIC METHODS

Water can be removed by: roller drying, spray drying, tunnel drying, fluidised bed drying, freeze drying, smoking and osmotic methods, utilising salt or sugar.

Salting is one of the oldest methods of preserving food. For more than a century, fishermen have preserved their catches by packing them in salt. When a micro-organism is in contact with a concentrated solution, e.g. a salt or sugar solution, water passes out of the micro-organism through the cell surface membrane by osmosis into the solution. As a result the cells of the micro-organism become dehydrated and incapable of multiplication, and the food is preserved.

Jams and other foods, e.g. condensed sweetened milk, preserved by sugar are protected from the growth of moulds by the high concentration of sugar. If the surface of the jam is exposed and water condenses there, the concentration of sugar at the surface may be reduced to a level at which moulds can grow.

12.4.2 TEMPERATURE CONTROL

REFRIGERATION

Refrigeration at 0–5 °C slows the rate of growth of micro-organisms.

When food is stored at 0–5°C, micro-organisms grow only slowly. Bacterial spores survive.

DEEP-FREEZING

Deep-freezing removes water from the state in which it can be used by micro-organisms.

Most fresh foods contain over 60% water. This water contains dissolved substances, and it is a property of aqueous solutions that the freezing temperature is below 0 °C. The lowering of the freezing temperature depends on the concentration of the solution. A solution containing one mole of solute per dm^3 freezes at -18 °C. At -5 °C, 60% of the water in peas is frozen; at -18 °C, 87% is frozen; at -30 °C, 92% of the water is frozen. If water is frozen, it cannot be used by micro-organisms. This is why at temperatures of -18 °C, the temperature of home freezers, food can be preserved for long periods. It is important to freeze foods quickly: the temperature at the centre of the food should pass from 0 °C to -4 °C within 30 minutes.

Commercial freezers are of a number of types:

Commercial freezers are of many types.

- Plate freezers: Food is packed on trays in a large cabinet which is cooled by a cold circulating liquid.
- Blast freezers: Food is placed in a cabinet or in a tunnel and frozen by a blast of cold air.
- Fluidised bed freezers: Food, e.g. peas, floats on a cushion of cold air as it passes through the freezer tunnel.
- Cryogenic freezing: A very cold liquid, e.g. liquid nitrogen, is sprayed on to the food. Freezing takes place very rapidly, and the food retains its original shape and appearance. The technique is expensive and has been used for fruits, e.g. raspberries and strawberries, and seafood, e.g. prawns and scampi.

Some vegetables and fruits need to be blanched before freezing [see under 'Freeze-drying' above].

When food is frozen, the water in it forms ice crystals. As it freezes, water expands and causes a partial breakdown of the cell walls. When food is thawed, the cell contents seep

In frozen and subsequently thawed foods the cell walls break down.

through the damaged cell walls. This is why thawed foods will keep less well than fresh foods. In the case of foods with a high water content, e.g. strawberries and tomatoes, the cellular breakdown on thawing is severe and the thawed foods are unappetising.

Frozen food loses little of its nutritional value. Commercial firms transport vegetables from the field to the freezer in a matter of hours. The nutritional value of frozen food which has been quickly frozen and well stored may be greater than that of 'fresh' food which may be eaten several days after it was harvested.

The nutritional value of frozen foods is high. Some nutrients may be lost in 'drip'.

When frozen foods are thawed, there is some loss of liquid – 'drip' – which carries away some of the soluble nutrients in the food. The extent to which thawed foods drip depends on the rate at which they were frozen, the time and temperature of storage and the nature of the food. When vegetables are thawed, there is some loss of vitamin C, and frozen vegetables are best cooked without thawing.

Some thawed fruits are soft and mushy. The liquid should be consumed with the fruits to avoid loss of taste and nutrients. Thawed meat may lose soluble proteins and vitamins in the liquid. Wastage of these nutrients can be avoided by making the liquid into a gravy.

12.4.3 STERILISATION

Sterilisation requires heating foods to a temperature which is high enough to kill micro-organisms and their spores.

Heating foods to a sufficiently high temperature kills micro-organisms and their spores. The process is called **sterilisation**. If the food is to be kept for a long time, it must be sterilised in a sealed container, e.g. a can or a bottle, which will prevent fresh micro-organisms from entering after the food has been sterilised. Nearly all foods can be canned or bottled. Some vegetables must be blanched first. Some foods, e.g. fruits, are mixed with sugar syrup in the can and heated to 95 °C before the can is closed.

Canned vegetables and meats are heated to 115 °C, fruits are heated to only 100 °C. They are more acidic than meat and vegetables, and this acidity protects them. The food is heated in the can, producing steam which drives out air, and the can is closed. As the contents cool, the reduction in pressure sucks in the ends of the can [see Figure 12.4(a)]. A bulge in the can means that gas has been formed from micro-organisms reproducing and respiring. The can is 'blown' and should be destroyed. For acidic foods, e.g. fruits, a lacquered can must be used to prevent acids from reacting with tin plate.

FIGURE 12.4A
(a) Canned food
(b) A 'blown' can

 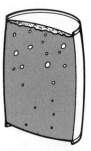

The process of canning involves heating foods to 115 °C – or 100 °C for fruits. Pressure cookers make sterilisation easier.

The HTST technique of holding at 121 °C for 3 minutes is called the botulinum cook.

The use of pressure cookers has made sterilisation methods much more efficient. Under a pressure greater than one atmosphere, the boiling temperature of water is higher than 100 °C, The modern HTST (high temperature short time) technique employs a temperature of 121 °C for 3 minutes. These conditions are drastic enough to kill spores of *Clostridium botulinum* which are particularly dangerous, and all other harmful organisms, and are described as the 'botulinum cook'; that is, the minimum safe process.

Properly canned food remains edible for many years, provided the cans are not corroded. Preserving meat and fish is too risky for home preserving methods. Preserving fruits and vegetables is less risky because the acids they contain make it more difficult for micro-organisms to survive, and home preserving methods are successful.

PASTEURISATION

Pasteurisation is treatment at 72 °C for 15 seconds. It preserves milk for several days.

Pasteurisation is partial sterilisation. The process was named after Louis Pasteur, the famous French scientist. The most important application of pasteurisation is in making cows' milk safe for human consumption. Pasteurisation kills many but not all the bacteria in the milk. Fortunately, the harmful bacteria are killed. The usual treatment of 72 °C, for 15 seconds enables milk to keep for several days if it is kept cool. Pasteurisation does not alter the taste of milk.

12.4.4 ATMOSPHERE CONTROL

Removal of oxygen from the space around foods restricts the growth of moulds and aerobic bacteria. Yeasts and anaerobic bacteria can still grow. The removal of oxygen is therefore used in conjunction with another preservation technique, e.g. freeze-drying and canning (see above). Some fruits and vegetables can be preserved by storage in a controlled atmosphere. By controlling the fraction of oxygen in the atmosphere in contact with the food, it is possible to delay the changes which take place during ripening and prolong the storage life of the food.

FIGURE 12.4A
Sealed under Nitrogen

Exclusion of oxygen restricts the growth of moulds and aerobic bacteria.

12.4.5 CONTROL OF pH

Micro-organisms grow more slowly in acidic conditions.

In acidic foods the growth of micro-organisms slows down. Vinegar (a 3% solution of ethanoic acid) is used in pickling. Yoghurt keeps well because it contains lactic acid which has been formed by the action of micro-organisms on lactose.

12.4.6 CURING

Some salts will inhibit the growth of micro-organisms. Meats, e.g. bacon, are cured by being soaked in or injected with a solution of sodium chloride, potassium nitrate and

Curing involves treating meats with sodium chloride, potassium nitrate and potassium nitrite.

sodium nitrite. There is concern over the use of nitrite in curing solutions because nitrite could react with secondary amino groups in protein foods to form nitrosoamines, which are carcinogenic. The danger of food poisoning from *Salmonella* is, however, much greater than the danger from nitrites.

12.4.7 SULPHITES AND BENZOATES

Sulphur dioxide and sulphites are used as preservatives. They act as reducing agents, inhibiting the autoxidation of carbonyl compounds and unsaturated fats in food. They restrict the growth of micro-organisms by removing oxygen from the environment and by creating acidic conditions. Foods which are preserved by sulphur dioxide and sulphites include sausages, dried fruits, soft drinks and many other items. Benzoic acid and benzoates are added as preservatives to soft drinks.

FIGURE 12.4B
Preserved by Sulphur Dioxide

Sulphites and benzoates act as preservatives through their reducing action.

FIGURE 12.4C
Preserved by Benzoic Acid

12.4.8 IRRADIATION

Irradiation of food by X-rays or γ-rays is effective in killing 99% of the micro-organisms which spoil food.

Irradiation of food means exposing food to X-rays or to γ-rays from radioactive elements. Very small doses of radiation are enough to kill 99% of the micro-organisms that spoil food, including *Salmonella* and *Clostridium*. All foods contain a natural low level of radioactivity. The dose of radiation which the food receives raises this level only slightly. By the time the food is eaten, the extra radioactivity has decayed.

The UK is one of the countries which bans the irradiation of food.

Many countries irradiate food, but in the UK the irradiation of food is at present banned. A committee set up by the Government advised in 1986 that the Government should allow the use of radiation for preserving food, but the recommendation has yet to be acted on. A poll of 7000 people in 1989 showed that 51% were against irradiating food.

There is some loss of vitamins A, B_1, B_6, C and E when food is irradiated. Potatoes are irradiated to stop them sprouting, without altering the taste. Some tests show that irradiation destroys 30–40% of a food's vitamin C, but other tests put the figure at 3–4%. Potatoes are an important source of vitamin C in the diet of the average person so the loss of some of their vitamin C may be serious.

A disadvantage is the loss of some vitamins when food is irradiated and the loss of texture and colour in some foods.

Irradiation does not work well for all foods. Tomatoes and strawberries become soft, lobsters turn black, eggs develop a smell, and red meats turn brown and develop a peculiar taste. It works well for spices. Being imported from tropical countries, spices are liable to contain micro-organisms. They can be irradiated with no loss of flavour.

Many people would benefit from the reduction in spoilage and increased shelf-life of irradiated food. Food manufacturers could deliver to retailers less frequently in larger batches, thus cutting distribution costs. Supermarkets and other shops could keep irradiated food on the shelves longer, thus reducing waste and being able to offer a wider variety of goods. Shoppers could keep food longer and shop less often.

The major benefit would be the reduction in food poisoning, but food prepared in unhygienic homes and restaurants could still be contaminated after sale. In countries where there is a danger of famine, irradiation could be used to protect food stores from moulds and bacteria. However, these countries cannot at present afford the technology for irradiating their food stores.

One concern is that a certain batch of food could be irradiated many times. Opinions on the benefits of food irradiation are divided.

It is possible to irradiate a food many times to prolong its life. A batch of, say, prawns, could be kept until it had a high bacteria count, then irradiated to kill the bacteria, stored again, irradiated again and stored again until needed for sale. Food in shops can be labelled [see Figure 12.4D] so that customers can decide whether or not they wish to buy it, but people eating in a canteen or restaurant cannot tell whether they are eating irradiated food. There is no test that will tell whether food has been irradiated. This seems to be an argument in favour of the safety of irradiated food!

In favour of the irradiation of food are the World Health Organisation, the UN Food and Agriculture Organisation and the UK Government Advisory Committee on Irradiation and Novel Foods. Against the irradiation of food are the International Organisation of Consumers Unions (170 organisations in 70 countries) and the London Food Commission. The UK Consumers' Association has expressed the fear that food irradiation may allow food manufacturers to be more lax over hygiene.

FIGURE 12.4D
Symbol for Irradiated Food

12.4.9 MILK

'Little Miss Muffet sat on a tuffet eating her curds and whey'. What makes milk turn into curds and whey? 'The milk has curdled' used to be a common complaint in the days before so many houses had refrigerators.

Fresh milk, straight from a healthy cow, contains few bacteria. As soon as it has left the udder, milk becomes contaminated by bacteria from soil and water on the exterior of the udder, from the hands of the milker and from the milking machines. By the time milk reaches the factory, it has a bacterial count of up to ten thousand million organisms per dm^3.

Lactobacilli convert lactose in milk into glucose and galactose and finally into lactic acid. The falling pH makes the proteins in milk separate as 'curd' from the liquid 'whey'.

Bacteria called *Lactobacilli* contain the enzyme lactase which converts the sugar lactose into glucose and galactose. This is the first step in a sequence of enzyme-catalysed reactions that produce lactic acid (2-hydroxypropanoic acid, $CH_3CHOHCO_2H$). As the pH of the milk falls with the production of lactic acid, the milk proteins, casein, globulins and albumins, are affected. Casein exists in association with calcium, magnesium, phosphate and citrate ions. As the pH falls, the bonds with calcium and phosphate are broken and casein becomes less soluble. At about pH 5, it separates out and coagulates, a process described as *curdling*. Albumins and globulins remain in the liquid phase as *whey*. Eventually the falling pH inactivates the bacteria which produce lactic acid so that no further lactic acid is produced. The product is known as *'curds and whey'*.

Pasteurised milk contains few Lactobacilli. *Sterilised milk has been held at 121 °C for 3 minutes.*

If many organisms other than *Lactobacilli* are present, spoilage does not stop at lactic acid and many other products form, e.g. ethanoic acid, ethanal, ethanol, propanone, carbon dioxide and others. Pasteurised milk (which has been kept at 72 °C for 15 seconds; see above) contains few *Lactobacilli* and takes longer to spoil than fresh milk. Spoilage involves protein breakdown and fat breakdown, and the products, which include amines and hydrogen sulphide, are more unpleasant than lactic acid.

UHT milk has been held at 132 °C for 2 seconds.

Sterilised milk has been treated at 121 °C for 3 minutes. UHT milk is made by the **ultra-high temperature treatment** (132 °C for 2 seconds) which kills all bacteria, while changing the taste of the milk only slightly. **Evaporated milk** is made by removing one third of the water before sterilising and canning the milk. **Condensed milk** is made by removing two-thirds of the water and adding sugar (45%) to preserve the milk. Condensed milk is unsterilised.

12.4.10 CHEESE

Cheese manufacturers speed up the formation of lactic acid which occurs naturally through bacterial action by adding rennin. The enzyme rennin is present in the extract called rennet which is obtained from the digestive juices in calves' stomachs. Rennin helps casein to separate from solution, and clotting occurs. The present demand for rennin is too high to satisfy from this source, and rennin is now synthesised by genetic engineering [see § 5.8]. Cottage cheese is unripened curd from skimmed milk. Cream cheese is unripened curd from thin cream.

Cheese makers add rennin to milk to speed up the formation of curd. Curd is ripened in various ways to form different cheeses.

Curd is ripened, treated in various ways, to give different types of cheese. The method of ripening decides which type of cheese is produced. Ripening is brought about by a large selection of bacteria and moulds, which are spread on the surface of the curd or injected into it or added to the milk before curd formation. Each micro-organism has a different set of enzymes, which catalyse a different set of reactions and result in the formation of a unique cheese. Some of the fat in cheese is converted into fatty acids which give distinctive tastes and smells to the different cheeses.

CHECKPOINT 12.4

1. Why is it necessary to use a pressure cooker when canning some foods?

2. What effect does acidity have on the survival of spores?

3. How does a high sugar content act to preserve food?

4. Why does meat from a well-fed rested animal keep better than meat from a hungry exercised animal? [see § 11.4].

5. What is the idea behind dehydration and freeze-drying?

6. What is the value of quick-freezing?

7. Describe the types of heat treatment used to preserve milk. What are the advantages and disadvantages of each method?

8. Why are the following used in the production of bacon? (*a*) sodium chloride, (*b*) sodium nitrite.

9. Food spoils because chemical reactions take place in it. How are some of these reactions inhibited by (*a*) cooking, (*b*) refrigeration, (*c*) freeze-drying?

10. The most popular method of preserving foods in terms of quantity and variety is canning. Explain the advantage of canning over other methods.

11. (*a*) What are the dangers posed by *Clostridium botulinum* in food?

(*b*) What is done to combat the bacterium?

12. (*a*) Name two micro-organisms which are responsible for food spoilage.

(*b*) Give an example of a beneficial change in food which is brought about by micro-organisms.

13. An apple is peeled and cut into segments. Say what you could do to stop the segments turning brown. Explain the basis of each method you suggest.

14. Draw a graph of the logarithm of the bacterial population in a contaminated food against time. State what each section of the graph represents.

15. Peas are preserved by canning, freezing and dehydration. Explain why each of these methods is successful in preserving peas. Compare the methods with respect to (*a*) cost, (*b*) long life, (*c*) taste.

16. Copy and complete the following summary:
The breakdown of carbohydrates by _____ gives rise to _____ acid, ethanol, ethanal, propanone, carbon dioxide and other products.
The breakdown of proteins by _____ gives peptides and _____ acids, the gases _____ and _____ and other products.
The breakdown of fats by _____ gives _____ acids which may be oxidised to long-chain _____ compounds.

12.5 ECONOMIC ASPECTS OF FOOD PROCESSING AND PRESERVATION

Some foods are eaten as they are harvested . . .

'I would prefer to live in a world where we harvested our foods fresh from the earth, ate them immediately and never gave a thought to food preservatives, artificial emulsifiers and stabilisers, anti-oxidants and permitted colours. Alas, we do not live in such a world. High technology food production and elaborate chains of food distribution have created a situation in which food additives are necessary.'

This is what Leslie Kenton wrote in her foreword to Maurice Hanssen's book, *E for Additives* (Thorsons, 1984). Interest among the public in the subject of food additives is so great that the book was a best-seller.

We eat some foods exactly as they are harvested, e.g. apples and lettuces. Most foods, however, go to the food industry to be processed, that is changed in some way to make them more appetising or to make them keep longer. Three-quarters of the food eaten in the UK is processed. You can eat potatoes fresh from the ground at a cost of 40 p per kilogram, and you can eat potato crisps, which are a product of food processing, at 600 p per kilogram.

. . . but most pass through the food industry for preservation and processing.

The different methods of preserving food vary in cost. Smoking fish and salting fish are less costly than deep-freezing, but the taste of frozen fish is much closer to that of fresh fish. Dried peas are less expensive than canned peas, which are less expensive than frozen peas, but frozen peas taste like fresh peas.

Food	*Price of one kilogram*/p			
	Dried	*Canned*	*Frozen*	*Fresh*
Green beans	—	100	110	460
Apricots	600	360	—	240
Cod	—	—	500	690
Salmon	—	370	610	1090
Beef	—	280	495	680
Potatoes	360	90	65	50

TABLE 12.5A
Comparing Food Prices

Foods must be transported from the producer to the food processor and from the factory to the wholesaler and then to the retailer. Most transport takes place by road, and the cost therefore depends on how tightly the food can be packed. A low-density load, such as a truck load of lettuces, does not cost much less to transport than a high-density load, such as a truck of canned meat. To transport frozen food, a refrigerated lorry must be used, and this increases the cost of frozen foods.

Foods which are to be transported must be packaged. Piling cauliflowers into a truck would get them from the market garden to the retailer, but when they arrived those at the bottom of the load would be crushed. The cauliflowers must therefore be packed into crates which are strong enough to protect the vegetables in them during transport. The more fragile the food item is, the more care and expense go into packing it for transport.

Transport costs add to the prices of food items.

QUESTIONS ON CHAPTER 12

1. (*a*) Name two pathogens which are responsible for food poisoning.

(*b*) Name two organisms which are not pathogens and which cause food spoilage.

(*c*) Give an example of a beneficial change in food which is brought about by bacteria.

(*d*) Sketch a typical growth curve to show the change in the bacterial population in a contaminated food sample.

2. Cut apples and potatoes turn brown on exposure to air. Describe experiments you could do to show that

(*a*) micro-organisms are not involved

(*b*) air is necessary for browning to occur

(*c*) the browning has the characteristics of an enzyme-catalysed reaction.

3. What is your opinion on the irradiation of food? Are the advantages great enough to make it advisable to admit food irradiation in the UK? Is there a risk which makes it unwise to allow food irradiation in the UK?

4. The table lists the prices of several kinds of milk.

Milk	*Price/* p	*Mass/* g	*Price/* (p kg^{-1})
Pasteurised	53	1140	
Sterilised	60	1000	
Condensed	89	405	
Evaporated	51	410	
Dried	106	198	

(*a*) Find the price per kilogram of each type of milk.

(*b*) Explain how the differences in price arise.

5. Explain the biochemical reasons for the following practices.

(*a*) eating wholemeal, rather than white bread [see § 11.2]

(*b*) adding enzymes, amylases, to flour in bread manufacture [see § 11.3]

(*c*) storing fruit in an atmosphere of carbon dioxide [see § 12.3]

13

FOOD ADDITIVES

Food additives are substances which are not normally present in a particular food and which are added for specific reasons. A food manufacturer must give a good reason for using an additive. This might be to preserve the food, to flavour it, to colour it or to alter its texture.

13.1 PRESERVATIVES

The spoilage of food by atmospheric oxidation has been mentioned [§ 12.2]. Foods containing fats become rancid on storage [see § 4.5]. Vitamin A, carotene and vitamin K are lost through oxidation. An **anti-oxidant** is any substance which can prevent or delay the deterioration of food due to oxidation. Vitamin E is one of the anti-oxidants which occur naturally in vegetable fats. The quantities present are not large enough to prevent oxidation over a long period. Common antioxidant additives are BHA (2-butyl-4-methoxyphenol) and BHT (2,6-dibutyl-4-methylphenol) [see § 4.5].

Anti-oxidants are added to delay the deterioration of foods through oxidation.

The use of sulphur dioxide and sulphites, benzoic acid and benzoates as additives was mentioned in § 12.4.

13.2 FLAVOURINGS

Flavourings are the largest group of additives, with 3000 or so in use. Some are natural products, extracted from plants, e.g. orange and peppermint. Some flavourings are synthetic substances; these are generally less costly than natural products. For flavours, see § 10.4. Examples of flavourings are

- ethyl methanoate, $HCO_2C_2H_5$, rum flavour
- propyl pentanoate $C_5H_{11}CO_2C_3H_7$, pineapple flavour

The commonest sweetener is sucrose. This is a food, not an additive. Many people want to reduce their intake of sucrose either because they are overweight or because it causes tooth decay. Additives which can be used as a substitute for sucrose include saccharin, sorbitol, mannitol and aspartame.

Flavourings are the largest group of additives. The flavour enhancer monosodium glutamate is widely used.

Flavour enhancers are not flavourings: they are substances which make existing flavours seem stronger by stimulating the taste buds. The commonest flavour enhancer is monosodium glutamate (MSG), $HO_2CCH(NH_2)CH_2CH_2CO_2Na$. Some people are allergic to MSG.

13.3 COLOURING MATTER

When food is processed it may lose some of its original colour; then the manufacturer will want to restore the original colour of the food. **Natural colouring materials** [see § 10.6] include carotenoids in carrots and tomatoes [Figure 10.6A], chlorophyll [Figure 10.6A] in green leafy vegetables and anthocyanins in plums and strawberries. Cochineal and saffron are natural colourings. **Synthetic colouring materials** are much less costly, and many coal tar dyes have been made for use in foods. Some of them came under suspicion of being carcinogenic, and some were associated with allergies, with the result that the number of permitted food colourings is tightly controlled by European Community (EC) regulations. Forty-six colouring additives are permitted in the UK. Some foods may not have colouring matter added, e.g. fresh meat and poultry, fish, raw fruits and vegetables, baby foods.

Synthetic colouring materials restore the appearance of foods after food processing.

13.4 TEXTURE CONTROLLERS

A large number of additives are used to alter the texture of foods. Chief among these are **emulsifiers**, which enable oil and water mix to form an emulsion [see § 4.14]. Margarine, ice-cream, salad dressings and many desserts contain emulsifiers. Lecithin [see § 4.14] is widely used as an emulsifier. A **stabiliser** is a substance which helps to prevent the emulsion from separating into oil and water. It protects the droplets from coalescing by increasing the viscosity of the medium. Examples are monohexadecanoylglycerol, carboxymethylcellulose, agar, alginic acid (from seaweed) and carob bean gum.

FIGURE 13.4A
All these contain emulsifiers and stabilisers

Anti-caking agents are added to powdered foods, e.g. cake mixes, and crystalline foods, e.g. table salt, to prevent lumps from forming. Many are anhydrous salts which can absorb water without becoming wet, e.g. magnesium hexadecanoate and calcium silicate.

Humectants are added to products such as bread and cakes to keep them moist. Examples are glycerol, mannitol and sorbitol.

Texture controllers include emulsifiers, stabilisers, anti-caking agents, humectants, thickeners and gelling agents.

Thickeners are added to many soups and puddings. Examples are corn sugar gum and modified starch, e.g. ethanoyl starch and phosphoryl starch.

Gelling agents, e.g. pectin, agar and calcium alginate, are added to jams and desserts to make them set.

13.5 CONTROLS ON ADDITIVES

Before a food additive may be used, it must be approved by the Government. The manufacturer then applies to have the additive licensed by the EC. If it is approved, by the EC, the additive is given an 'E number' – the letter E followed by a number. The numbering system is:

Colouring matter: E number begins with 1, e.g. E150 caramel

Preservatives: E number begins with 2, e.g. E221 sodium sulphite

Anti-oxidants: E number begins with 3, e.g. E330 citric acid

Texture controllers: E number begins with 4, e.g. E461 methylcellulose.

Additives that have been passed in the UK but not yet in the EC have a number only, e.g. 107 yellow 2G, 524 sodium hydroxide, 925 chlorine, 621 MSG.

Additives are tested on animals before they are licensed. Some additives, however, produce **intolerance reactions** in some people. These intolerance reactions or **allergies** can take the form of asthma, eczema, rhinitis (like hay fever), headaches, migraines, digestive troubles and hyperactivity. Many hyperactive children improve dramatically when they are put on a diet free from tartrazine E102, a yellow dye which is used in sweets, fizzy drinks and desserts. The labels on foods must contain a list of the ingredients, including additives with their E numbers. This means that someone who has an allergy to an additive, e.g. tartrazine, can reject any foods which contain it.

Food additives approved by the EC are allocated E numbers. Some additives and some natural foods provoke intolerance reactions – allergic reactions – in some people. Guidelines on food additives have been drawn up by the WHO.

The guidelines drawn up by the World Health Organisation are that the use of food additives is justified when they serve one or more of the following functions:

* to maintain the nutritional quality of food
* to make food keep longer
* to make food more attractive, provided that the consumer is not deceived.
* to provide essential aids to food processing.

QUESTIONS ON CHAPTER 13

1. Cooked ham may contain E250 sodium nitrate (up to $500 \, \text{mg kg}^{-1}$) and E450 sodium heptaoxodiphosphate(v) (sodium pyrophosphate). The function of the sodium heptaoxophosphate(v) is to act as an emulsifier to retain water that has been injected into the ham.

(*a*) What is the function of the sodium nitrite in the ham (see §12.4.6 if necessary)? Who benefits from the addition of sodium nitrite to ham?

(*b*) Who benefits from the addition of sodium heptaoxo-phosphate(v) to ham?

(*c*) Comment on the relative merits of the two additives.

2.

> **CHEESE SPREAD**
>
> Ingredients: Cheddar cheese, Skimmed Milk,
> Whey powder, Butterfat,
> Emulsifiers E450a, E450b, E450c, Salt, Colour E160a,
> Preservative E202, Anti-oxidant E320

Refer to the label on a packet of cheese spread.

(*a*) What is the function of (i) the preservative E202 (potassium sorbate), (ii) the anti-oxidant E320 (BHA), (iii) the colouring agent E160 a (carotene), (iv) the emulsifiers E450 (various sodium phosphates)?

(*b*) Why do emulsifiers feature so largely on the label?

3. Refer to the label on a packet of fish fingers.

> **Fish fingers**
>
> Ingredients: Cod fillet, Breadcrumbs,
> Colours E102, E120,
> Batter with flavour enhancer 621,
> Hydrogenated vegetable fat, Emulsifier E450c

(*a*) The flavour enhancer here is MSG. What is a flavour enhancer? What does MSG stand for? Is there any disadvantage of the use of MSG?

(*b*) What is the function of an emulsifier? How does the molecular structure of an emulsifier enable it to perform this function?

(*c*) What is meant by 'hydrogenated vegetable fat'? What is the advantage of hydrogenation?

(*d*) The colouring substances E102 and E120 are tartrazine and cochineal. What purpose do they serve? Is there any disadvantage to using either of them?

14

FOOD LEGISLATION

Before the industrial revolution, most people grew their own food or bought food from farmers in their community. Food legislation was not needed because people could not adulterate food which they supplied to their neighbours or regular customers. Early in the nineteenth century, industrial towns were growing rapidly, and their populations had to be supplied with food. The suppliers were remote from the consumers, and opportunities to cheat arose. For example, thickening agents might be added to cream, water might be added to spirits and fruit juices, and harmful colouring agents might be added to make food items look more appetising. The adulteration of food became so bad that Parliament passed an act to prevent it.

The need for food legislation came after the industrial revolution and the movement away from the countryside.

14.1 FOOD ADULTERATION ACT 1860

The **Food Adulteration Act** was passed in 1860. Its aim was to protect the public against the addition of other substances to foods. Analysts were appointed to detect adulteration of food. Fraudsters then faced the possibility of being prosecuted if their wrong practices were discovered. In 1860, analytical chemistry was not as advanced as it is today. The techniques of paper chromatography, thin layer chromatography, gas liquid chromatography and spectroscopy had not been developed. The adulterants were therefore difficult to detect, and the law was difficult to enforce.

The 1860 Act aimed to protect the public from the adulteration of food.

14.2 FOOD AND DRUGS ACT 1955

A number of laws were passed between 1860 and 1955. In 1955 the laws were brought together into the Food and Drugs Act 1955. The five important principles of the Act are:

1. Food should be fit for human consumption and free from health hazards. For example, a sausage roll must not be mouldy and it must not contain any foreign matter.

2. Food must be of the nature, substance and quality demanded. For example, a bottle of spirits must contain the correct percentage of alcohol. An orange flavour drink cannot be sold as pure orange juice.

3. Premises where food is prepared, that is, manufactured, baked or canned, must be controlled. Inspectors visit food manufacturers' premises form time to time.

4. Premises where food is sold, e.g. shops, vending machines, stalls, etc., must be controlled.

The 1955 Act laid down five principles governing the quality and safety of foods. The improvements in analytical techniques made it possible to enforce the 1955 Act.

5. Each local authority must appoint a fully trained public analyst with responsibility for analysing samples sent in for analysis by sampling officers and by members of the public.

Anyone who has a complaint about the quality of food which they have bought in a shop or a restaurant can report the matter to the local environmental health officer. He or she will then take action to have the public analyst analyse the food and will prosecute the supplier if necessary. Committees which keep the safety of our food under review and advise the Government are:

Food Standards Committee – advises on composition and labelling of food
Committee on Medical Aspects of Food – advises on health aspects
Food Additives and Contaminants Committee – advises on additives and on methods of preservation

14.3 FOOD LABELLING REGULATIONS 1980

In the European Economic Community (EC) food must be able to circulate freely between the member states. To overcome differences in national food standards, the EC brought out a set of directives on food which all member nations must obey. In 1980, the UK passed a new Food Labelling Regulations Act, which came into force in 1983. The Act made the food industry give more information on food labels.

FIGURE 14.3A
Parts of some Food
Labels

The 1980 Act made the food industry give more information on labels:

... the name of the food, a list of the ingredients, the net weight, the 'best before' date, the name of the manufacturer, instructions for use.

Some manufacturers also list the nutritional ingredients.

14.3.1   FOOD NAMES

All prepacked foods and almost all non-prepacked foods must show the name of
the food, e.g. pasta, cornflakes, lentils. If a manufacturer gives a food product a
trade name, then a description must be added to tell the customer what the
product is, e.g. Apricot Delight – *An apricot flavour dessert mix* and Pasta Paolo –
Pasta ribbons in a creamy Parmesan sauce. The name must not mislead the
customer. If the flavour of a yoghurt comes mainly from real strawberries, the
yoghurt can be called *strawberry yoghurt* or *strawberry flavoured yoghurt.* If the
strawberry taste is due to an artificial flavour, the yoghurt can be called *strawberry
flavour yoghurt.*

14.3.2   LIST OF INGREDIENTS

The labels on prepacked foods must list all the ingredients in order of decreasing
weight. Food additives are included. The labels on some foods do not have to list
the ingredients; these include fresh fruit and vegetables, cheese, butter and
yoghurt.

FIGURE 14.3B
Read the ingredients

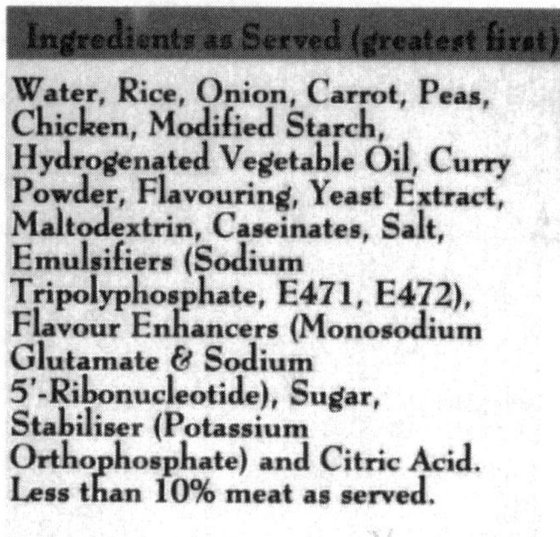

> **Ingredients as Served (greatest first)**
>
> Water, Rice, Onion, Carrot, Peas,
> Chicken, Modified Starch,
> Hydrogenated Vegetable Oil, Curry
> Powder, Flavouring, Yeast Extract,
> Maltodextrin, Caseinates, Salt,
> Emulsifiers (Sodium
> Tripolyphosphate, E471, E472),
> Flavour Enhancers (Monosodium
> Glutamate & Sodium
> 5'-Ribonucleotide), Sugar,
> Stabiliser (Potassium
> Orthophosphate) and Citric Acid.
> Less than 10% meat as served.

14.3.3   NET QUANTITY

The net weight (that is, without the packet) or the net volume of food must be shown
on the packet.

14.3.4   DATE MARKS

Most foods must be date-marked. The date mark usually consists of the words 'best
before' followed by the date until which the manufacturer expects the food to remain
at its best. Foods which keep for more than six months are excepted from date
marks. Examples are frozen vegetables and bottles of wine and spirits. Foods which
have a shelf-life of less than one month, e.g. yoghurts, meat pies, have a 'sell by' date.
Foods which are eaten soon after purchase are exempt; examples are apples and
cucumbers.

14.3.5 NAME OF MANUFACTURER

The name and address of the firm which manufactures or packs or sells the product must be given. The responsibility for the condition of the food lies with the people named.

14.3.6 INSTRUCTION FOR USE

Some foods would be difficult to use without instructions, e.g. cake mixes and dessert mixes. In these cases, the labels carry instructions for use.

FIGURE 14.3C
Instructions for Use

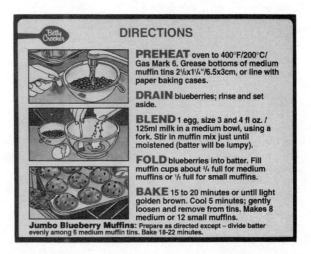

14.3.7 NUTRITION LABELS

It is not compulsory to list the nutritional ingredients of a food, but many manufacturers do this. The quantities of protein, fat and carbohydrate per 100 g of the food, the vitamin content and the calorific value of the food may be given.

FIGURE 14.3D
Is it good for you?

NUTRITION	100g serving gives you	50g serving gives you
Energy	1875kJ/ 450 kcal	938kJ/ 225 kcal
Protein	8.1g	4.1g
Carbohydrate	65.6g	32.8g
(of which Sugars	25.6g	12.8g)
Fat	16.9g	8.5g
(of which Saturates	11.3g	5.7g)
Fibre	4.5g	2.3g
(of which Soluble	1.7g	0.9g)
Sodium	0.04g	0.02g

Harvest Crunch contains no artificial colours, flavours or preservatives.

QUESTIONS ON CHAPTER 14

1. Explain the difference (if any) between (*a*) blackcurrant mousse, (*b*) blackcurrant flavoured mousse and (*c*) blackcurrant flavour mousse.

2. What information is missing from this label?

```
┌─────────────────────────────────────────────┐
│               CHEF'S DELIGHT                  │
│  ...............................................  │
├─────────────────────────────────────────────┤
│  Ingredients:  Soya protein, Dried vegetables, Modified │
│   starch, Flavourings, Salt, Colour E103, Emulsifier E322 │
│                                               │
│  _____    _____    _____       │
│  ...........    ...........    ...........     │
│  _____    _____    _____       │
├─────────────────────────────────────────────┤
│  Serving instructions                         │
│  Empty the contents of the packet into a saucepan. Add │
│  575 cm³ (1 pint) of cold water. Bring to the boil. Simmer for 10 │
│  minutes.                                     │
└─────────────────────────────────────────────┘
```

3. A manufacturer brings out a new cake mix. Design a packet which shows all the necessary information.

4. (*a*) Why do food manufacturers use additives?

(*b*) Why is the use of food additives subject to legal control?

(*c*) Give three examples of food additives and the reasons for their use.

5. How does the 1955 Food and Drugs Act protect the customer?

15

HUNGER

15.1 STARVATION

Six hundred million people are undernourished. Fifteen million children die each year from starvation and disease.

Every two seconds, a child dies of starvation or an illness caused by undernourishment. According to the World Bank, about 600 million people (12% of the world's population) are undernourished, and 15 million children die each year from starvation and disease. You have seen pictures on your television screen of children suffering from the extreme form of starvation called **marasmus**. They look like skeletons covered with skin. The children with swollen abdomens are suffering from **kwashiorkor**; they are getting some food but are starved of protein.

Why are these people starving? It is not because there is insufficient agricultural land in the world to feed all the world's population. Enough food is grown to provide each person in the world with more food than they need. It is in Third World countries that starvation and malnutrition are most common.

Undernourishment is more widespread in the developing countries, the Third World countries.

The world population is growing at a rate of about 2% per year. This adds 80 million people a year to the population. This is an average figure. In some European countries, e.g. Germany, the population is not increasing. In some African and Asian and South American countries the population is growing at a greater rate. In Nigeria, the population is increasing at over 3% a year, while the production of food is rising by only 0.5% a year, and food is imported to fill the deficit. Poorer countries cannot afford to import food, and their populations are undernourished.

The countries in the northern continents are richer than the countries in the southern continents – except for Australasia [see Figure 15.1A]. The poorer southern countries are called **Third World countries**. They are also described as **developing countries** because they are only now developing modern methods of agriculture and manufacturing industries.

In the past, many of these countries were under colonial rule and were not encouraged to develop industries.

Most of the Third World was at one time colonised by one or other of the western European nations. In areas which were suitable for growing tropical crops, the European colonists started plantations of sugar, tobacco, cotton, coffee and other crops. The plantations were owned and managed by Europeans, and the crops were exported to Europe. The profits from the plantations were invested in Europe, not in the colonies. While the colonies developed their ability to provide raw materials for export, they did not develop manufacturing industries of their own.

Colonial days are now over, and Third World countries have their independence. Their patterns of trade are still difficult to change, however. They still export raw materials – minerals and crops – and import manufactured goods. The crops which they grow for export and not for home consumption are called **cash crops**. Tea, sugar,

FIGURE 15.1A
This map of the world
shows the dividing line
between the rich countries
and the poor countries –
the Third World countries

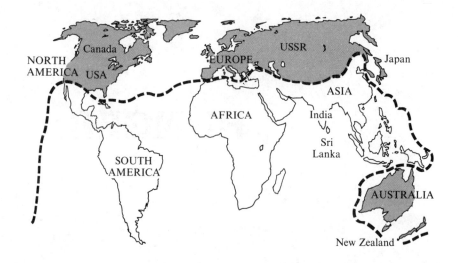

*Third World countries
devote land to growing
crops for export – cash
crops. Often these do not
bring enough income to pay
for the importation of
industrial equipment.*

coffee, cotton, cocoa, peanuts, bananas and many other foods are grown as cash crops. The problem is that since 1945 the prices of manufactured goods have risen much more steeply than the prices of raw materials. Third World countries have to export more to pay for their imports. Many Third World countries which want to develop and cannot afford to import the industrial equipment they need have borrowed from richer countries. They have run up enormous debts, and therefore need to export more to pay the interest on their debts. The Sudan spends 80% of its earnings from exports on paying interest on its debts of $11 billion. Brazil has debts of $100 billion. Because their incomes depend on trade, Third World countries do not want to cut down exports of cash crops. If land is being used to grow cash crops, it is not being used to grow food to feed the population.

15.2 SOME CASH CROPS

15.2.1 PEANUTS

The Sahel region of Africa [see Figure 15.2A] frequently suffers from drought. While Mali suffered a catastrophic drought in 1974, the country actually increased its exports of peanuts. Thousands of people starved while peanuts, which are a rich source of protein and could have saved lives, were exported. Most of the peanut crop is used to feed pigs and cows in Europe. European cows produce a surplus of milk which may be dried and sent to Africa for malnourished children.

15.2.2 TEA

About one million tonnes of tea are traded every year. Tea is dried and processed in the countries which grow it, so about half the price of the tea goes to the growers. The other half goes to the trading firms which package it, transport it and sell it. Tea-pickers live in poor conditions, often in barracks attached to the plantations. The pay for a woman is about 15 p a day in Bangladesh, 30 p a day in Sri Lanka and 45 p a day in Kenya. Men earn more and children earn less. Many of the children are undernourished, and many die in infancy. Trading companies claim that they must make a profit for their shareholders. In the UK, we drink an average of 6 cups of tea a day, which mounts up to about 4.5 kg a year. It costs us about 1 p per cup. If we paid

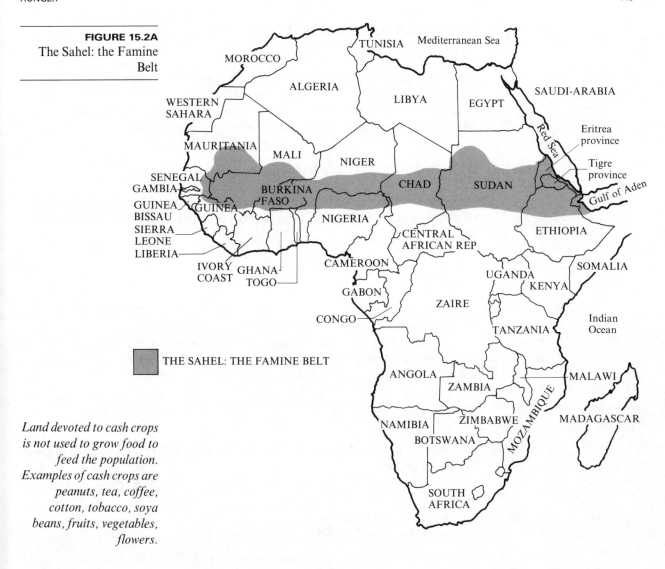

FIGURE 15.2A
The Sahel: the Famine Belt

THE SAHEL: THE FAMINE BELT

Land devoted to cash crops is not used to grow food to feed the population. Examples of cash crops are peanuts, tea, coffee, cotton, tobacco, soya beans, fruits, vegetables, flowers.

a little more or if shareholders were to make do with lower profits, tea pickers could be paid enough to provide them with decent living conditions.

15.2.3 COFFEE

In Ethiopia, the best land is used to grow coffee. Ethiopia is in the Sahel belt and has suffered two recent famines, in 1984–5 and 1987–8. During the 1984–5 famine, Ethiopia continued to export fruit and vegetables to the UK and other countries while several hundred thousand people starved. During the 1987–8 famine, bands of guerillas from the northern provinces of Eritrea and Tigre were fighting the Ethiopian Government, and the government kept up exports of food to pay for its well-equipped army. If the income were not needed for arms, the fertile land used to grow coffee could be used to grow food crops for the population.

Brazil uses large areas of good land to grow coffee for export, and this country has a large number of undernourished people. In a drive to finance industrial development, Brazil has found a way of reducing its expenditure on imported oil. Large areas of land are used to grow sugar cane, and the sugar is fermented to give ethanol which is used to replace petrol as a vehicle fuel. This project uses land which could otherwise be used for food production.

15.2.4 COTTON, TOBACCO, SOYA BEANS

Zimbabwe in Africa [see Figure 15.2A] suffered from drought in 1984. It had to import half a million tonnes of maize to feed people. At the same time, Zimbabwe was exporting cash crops of cotton, tobacco and soya beans. The country needs an income to support its large army. Would it not be wiser to become self-sufficient in food before devoting so much land to cash crops?

15.2.5 STRAWBERRIES, ASPARAGUS

Kenya [see Figure 15.2A] also suffered from drought in 1984. The areas with the best rainfall were in use for growing the cash crops, including strawberries, asparagus and other luxuries.

15.2.6 BANANAS

The growers receive only 12% of the price of a banana. The rest goes to exporters and importers, wholesalers and retailers. In 1960, a Caribbean country could buy a tractor with the income from 3 tonnes of bananas. In 1970, it took 11 tonnes and in 1990 it took 25 tonnes of bananas to buy a tractor. A Third World country cannot increase the price of its bananas because the customer can buy elsewhere.

15.2.7 FLOWERS

In Colombia in South America, malnutrition is common. Fertile land is used to grow flowers for export.

15.3 AID

Aid takes a number of forms:

food sent to famine areas, equipment and expertise to help with long-term development.

Developed countries help developing countries through direct aid. This may be food sent to areas stricken by famine. It may be money, equipment and trained staff to help with long-term technological development. Sadly, much of the food sent to famine areas never reaches the hungry people. It may be stolen or diverted to officials or held up by warfare or spoil en route because of inefficient transport. In the long run, the provision of the technology and training which will help a country to become self-sufficient in food is more productive.

The biotechnology involved in culturing algae, fungi and bacteria to provide protein foods was described in § 2.19. Mycoprotein (from fungi) is suitable for human consumption. The protein produced from bacteria is suitable for feeding animals, and further advances may produce a bacterial protein suitable for human consumption. Animals can be fed on bacterial protein to produce meat, milk and eggs for human consumption. The production of single cell protein is not affected by the weather and

The culturing of algae, fungi and bacteria as sources of protein could help to feed people who lack agricultural land.

requires a fraction of the land needed to produce vegetable protein. One million tonnes of Pruteen could be produced on a land area of 100 hectares, whereas 2 million hectares are needed to produce the mass of soya bean meal containing the same amount of protein. The technology exists, but so far there has been no transfer to developing countries.

The Brandt Commission, an international committee chaired by Herr Willy Brandt of Germany, reported in 1979 that the cost of a food relief programme for one year would fund a technological investment programme for five years. However,

The Brandt Commission of 1979 advised investment in long-term technological programmes in developing countries. Little has been done.

emergencies are constantly arising, and direct food aid is necessary to save lives. Herr Brandt commented, 'Every minute of every day of the week, the nations of the world are spending about $2 million on armaments and military equipment. And every minute 30 children under five years of age are dying from malnutrition.' The Brandt Commission reported in 1979, but the situation is the same today. The report received the approval of governments when it was published, but the same governments took very little action to put into effect the recommendations of the Commission.

15.4 TRADE

Third World countries export, to a large extent, raw materials. Developed countries export manufactured goods which bring higher profit margins.

The poor countries need more control over the price of their exports. They also need to be able to obtain higher prices for their products by processing raw materials themselves. Sri Lanka and Kenya export tea leaves to the UK. The UK exports tea bags and packaged tea, which fetch a higher price than loose tea. When Third World countries try to export manufactured goods, they often meet obstacles. The rich countries do not place limits on imports of raw materials, but they often place quotas and tariffs on imported goods. A quota is a limit on the value of imported goods. A tariff is a tax on an imported item. The removal of such restrictions would help Third World countries to increase their export trade. If they can export more, they will be able to import more, and trade will improve worldwide.

QUESTIONS ON CHAPTER 15

1. How can agricultural science

(a) increase the production of food in a developing country,

(b) reduce waste of crops?

2. The EC produces a surplus of food. Some African countries lack food. Moira thinks that tonnes of food should be shipped from the EC food stores to countries whose populations have an inadequate diet. Gwynneth thinks that this would stop African countries from working out a solution to the problems of drought and soil erosion. Comment briefly on each point of view.

3. In the 1987–8 famine in Ethiopia, that country exported beef to the UK. Some supermarket chains refused to buy the meat. Do you think this gesture would benefit Ethiopia

(a) in the short term and (b) in the long term?

4. Why are many people in Third World countries poor? Choose two statements with which you agree and two with which you disagree. Explain your choice.

● The birth rates are too high.

● The governments spend their income on the wrong things.

● The old colonial countries have taken the wealth of the Third World countries.

● The people are lazy.

● The people are not well educated.

● Banks and businesses make big profits out of poor countries.

● The climate works against them.

● Power is concentrated in the hands of too few people.

5. The supply of food to the hungry can be increased by

(a) increasing the amount of food produced or

(b) redistributing the food which the world already produces.

Which of these alternatives do you believe to be the better option? Explain your view.

6. Is it best for a Third World country which has a food shortage to invest in

(a) increasing food production or

(b) reducing food wastage or

(c) developing new food sources?

Say which of these choices offers (i) the best short-term solution and (ii) the best long-term solution.

7. The world produces enough grain to provide every person on Earth with 12 000 kJ of energy per day. In addition, vegetables and fruits are grown.
Why is it that, in spite of this production, one in eight of the world's population is severely undernourished?

8. Imagine you are empowered to make decisions in a Third World country which needs capital for industrial expansion and which also has a malnourished population. Would you devote land to cash crops or would you make growing crops to feed the people your priority? Explain your point of view.

9. Refer to §2.19 on the use of algae, fungi and bacteria as sources of protein and to the comment in §15.3.
Discuss whether use could be made of these techniques in Third World countries. Consider technological, social, educational, economic and political factors.

ANSWERS TO SELECTED QUESTIONS

PART 1: BIOCHEMISTRY

CHAPTER 2: PROTEINS

Checkpoint 2.5
1. (b) increase (d) increase
   (e) Interaction between side-chains will alter with ionisation; see §2.8.
2.
$$CH_2CO_2H$$
$$H_2NCHCONHCHCO_2CH_3$$
$$CH_2C_6H_5$$

Checkpoint 2.10
1. (a) See §2.1 (b) §2.8 (c) §2.2 (d) §2.8 (e) §2.7
2. (a) See §2.8
   (b) $-CO_2^-$ groups buffer against H^+; $-NH_3^+$ groups buffer against OH^-.
   (c) The zero overall charge reduces hydration.
   (d) The zwitterion structure; see §2.1.
   (e) In keratins in hooves, there is more cross-linking between neighbouring chains by e.g. disulphide bridges..
   (f) The disulphide links can be reduced to —SH groups, the hair shaped in the desired curls and the —SH groups then reduced to reform disulphide bridges.
3. (a) disulphide bridge, peptide link

   (b)
$$CH_3 \qquad CH_2CO_2H \quad CH_2OH$$
$$H_2NCHCONHCHCONHCHCONHCHCO_2H$$
$$CH_2$$
$$S$$
$$S$$
$$CH_2$$
$$H_2NCHCONHCHCONHCH_2CONHCHCO_2H$$
$$CH(CH_3)_2 \qquad\qquad CH_3$$

Checkpoint 2.14
2. (a) DFP inactivates the enzyme.
   (b) hydrogen bonding from the —OH group
3. (a) (i) 1 (ii) 0
   (b) The active sites are all occupied by substrate molecules.
   (c) limiting velocity or maximum velocity
   (d) Michaelis constant

Checkpoint 2.17
1. b
2. (a) See §2.16
   (b) 1. pH: affects ionisation of groups at the active site.
       2. Temperature: At first a rise in temperature increases the speed of the reaction, but a further rise denatures the enzyme.
       3. Cofactors; see §2.16
       4. Inhibitors; see §2.15
3. (a) See §2.13 (b) §2.13 (c) §2.15 (d) §2.11
4. (a) See §2.14 (b) §2.13 (c) §2.15
5. See §2.13

Checkpoint 2.18
1. The cost of extraction and purification is high.
2. The reaction takes place more slowly.
3. (a) See §2.18
   (b) The enzyme on its support can be either used in a batch process and then separated from the product and re-used or it can be used in a continuous process.
4. (a) See §2.18 and Figure 2.18A
   (b) See Table 2.17A. These processes can be run continuously with immobilised enzymes.

Questions on Chapter 2
1. (a) (i) [Enzyme] [Substrate] (ii) [Enzyme]
   (b) temperature and pH
   (c) the amount of product formed after measured intervals of time
   (d) See Figures 2.15B and C
   (e) e.g. chymotrypsin, peptide, DFP
2. For (a), (b), (c), (d), see §2.16
   (e) See §2.11 (f) See §2.13
   (g) an enzyme which catalyses isomerisation
3. Similarity: decrease in activation energy
   Differences: Enzyme-catalysed reactions are characterised by Michaelis–Menten kinetics, specificity, effects of temperature, cofactors and inhibitors.
4. (a) See §2.11
   (b) Specific activity of pyruvate decarboxylase
   = (Amount of CO_2/mol)/[(Time/s) (Mass of enzyme/mg)]
   For a known mass of enzyme, measure the rate of evolution of CO_2 by e.g. collecting in a gas syringe.
   For (c), (d) and (e) see §2.18

CHAPTER 3: CARBOHYDRATES

Questions on Chapter 3

1. (a) A, C   (b) B   (c) C   (d) A   (e) B
2. See §3.2
3. a, b, c, e
4. Potato crisps contain starch, whose molecules consist of glucose units joined by α-1,4-glycosidic links. These are hydrolysed by enzymes and the glucose is metabolised. Celery contains cellulose whose molecules consist of glucose units joined by β-1,4-glycosidic links. Humans do not possess enzymes which can hydrolyse this linkage, and cellulose passes through the alimentary canal without being digested; see §3.5.
5. (a) Starches have branched structures. Cellulose has a linear structure. Starches have α-1,4-glycosidic linkages, but cellulose has β-1,4-glycosidic linkages.
   (b) Humans do not have enzymes that can hydrolyse β-1,4-glycosidic linkages.
6. Maltose is reducing; sucrose is non-reducing. In sucrose the two anomeric carbon atoms are linked by the glycosidic linkage; there is no group

which is in equilibrium with

and can exert a reducing action [see §3.3].

7. Starch gives a blue colour with iodine; glucose and sucrose do not. Glucose gives a positive result with Benedict's reagent or Fehling's solution; sucrose gives a negative result.
8. (a) sugars and starches
   (b) oxidation to provide energy
   (c) glucose (α-D-glucopyranose), fructose (β-D-fructo-furanose)
   (d) sucrose

(The structure is shown also in §3.3).

CHAPTER 4: LIPIDS

Checkpoint 4.4

1. See §4.3
2. (a) See §4.1   (b) See §4.2
   (c) As in §4.4, reflux with a solution of KOH in ethanol
   (d) methanol + concentrated sulphuric acid
   (e) Separate as for amino acids in §2.6.2. Examine the chromatogram under UV light. Identify the esters by their R_F values.
3. 125
4. (a)

   $CH_2OCO(CH_2)_{14}CH_3$

   $CHOCO(CH_2)_{16}CH_3$

   $CH_2OCO(CH_2)_7CH=CHCH_2CH=CH(CH_2)_4CH_3$

   Any one of the three acids could esterify the middle CHOH of glycerol.
   (b) solid
   (c)

   $CH_2OCOC_{15}H_{31}$ $CH_2OH + C_{15}H_{31}CO_2Na$

   $CHOCOC_{17}H_{35} + 3NaOH(aq) \longrightarrow CHOH + C_{17}H_{35}CO_2Na$

   $CH_2OCOC_{17}H_{31}$ $CH_2OH + C_{17}H_{31}CO_2Na$

Checkpoint 4.10

For 1. and 2. see §4.5
3. (a) See §4.5   (b) See §4.7
   (c) Indicates degree of unsaturation

4. Amount of triglyceride $= \frac{1}{3} \times$ Amount of NaOH
   $= 1.5 \times 10^{-3}$ mol
   Amount of double bonds $= 2.5 \times 10^{-3}$ mol
   Degree of unsaturation $= 2.5/1.5 = 1.7$
5. For (a), (b) and (c) see §4.7
   (d)

   (e) $6.7\,cm^3$ thio, 6.7×10^{-4} mol thio, 3.35×10^{-4} mol I_2, 85.1 g I_2. Iodine value = 85.1
6. See §4.10
7. Growth of stems is promoted to such an extent that seedlings outgrow their strength; see §4.10

Checkpoint 4.12

1. (a) A = non-polar chain of phosphoglyceride/phospholipid molecule, B = polar head of phosphoglyceride molecule, C = integral protein/channel protein, D = carbohydrate chain of glycoprotein, E = peripheral protein
   (b) $W = 7$–8 nm   (c) See §4.12
2. The non-polar tail of a soap ion, being lipophilic, is attracted to the lipid part of a phospholipid, while the polar head is hydrophilic – attracted to water. Soap therefore has an emulsifying action on a phospholipid, e.g. a bacterial cell wall.

3. (a) See §4.12, including Figure 4.12C
   (b) See §4.11, including Figure 4.11A
   (c) See §4.12
   (d) glycoproteins; see §4.12
   (e) See §4.12 on active transport, including Figure 4.12D

Questions on Chapter 4
2. hydrolysis of lipids (§4.3) and autoxidation (§4.5)
3. (a) Hydrolysis is splitting a compound into two compounds by reaction with water. Hydrogenation is the addition of hydrogen across a carbon–carbon multiple bond.
   (b) Saponification is the hydrolysis of an ester to form an alcohol and the salt of an acid.
   (c) A phosphoglyceride is a mixed ester of glycerol with fatty acids and with an ester of phosphoric acid. A mixed triglyceride is an ester of glycerol with more than one fatty acid.
   (d) A fat is solid at room temperature; an oil is liquid at room temperature.
   (e) A wax is an ester of a high molar mass alcohol with a fatty acid.

4. 1-hexadecanoyl-2-octadecanoyl-3-octadec-9-enoylglycerol

$$CH_2OCO(CH_2)_{14}CH_3$$
$$|$$
$$CHOCO(CH_2)_{16}CH_3 \qquad + 3KOH(aq) \longrightarrow$$
$$|$$
$$CH_2OCO(CH_2)_7CH = CH(CH_2)_7CH_3$$

$$CH_2OH + CH_3(CH_2)_{14}CO_2K$$
$$|$$
$$CHOH + CH_3(CH_2)_{16}CO_2K$$
$$|$$
$$CH_2OH + CH_3(CH_2)_7CH = CH(CH_2)_7CO_2K$$

5. Soft margarine 40 °C, hard margarine 46 °C; The hard margarine contains more saturated lipids.
6. (a) the four-ring structure characteristic of steroids
   (b) Oestrone has a phenolic —OH; testosterone has an alcoholic —OH.
   (c) cholesterol
   (d) assistance in the digestion of lipids

CHAPTER 5: NUCLEIC ACIDS

Checkpoint 5.2
1. C, H, O, N – adenosine – ribose – five – purine – oxygen – cytosine – hydrogen
2. (a) (i) adenine, guanine (ii) cytosine, thymine
   (b) See §5.2, including Figures 5.2D and E
   (c) See §5.2. The sequence of bases in one strand of DNA can form hydrogen bonds to the sequence of bases in the other strand of DNA.
   (d) hydrogen bonding as shown in Figures 5.2B and C

Checkpoint 5.5
1. A codon is a triplet of bases in mRNA. An anticodon is the complementary triplet of bases in tRNA.
2. Transcription (see Figure 5.3B) is the assembly of a molecule of mRNA on a single strand of DNA. The genetic information in a strand of DNA is transcribed (copied) on to a molecule of mRNA. Translation (Figure 5.4A) is the assembly of a polypeptide chain on a molecule of mRNA. The information in the base sequence of mRNA is translated into a sequence of amino acids.
3. (a) See §5.5, including Figure 5.5A
   (b) the Meselson–Stahl experiments; see §5.5
4. (a) guanine, adenine, cytosine, thymine
   (b) See Figure 5.2B
   (c) See §5.3, including Figure 5.3B
   (d) See Figure 5.4A. Amino acids cannot bond to mRNA; tRNA can bond to an amino acid at one end and to mRNA at the other end.
5. (a) a sugar (ribose or deoxyribose) + a base (a purine or a pyrimidine) + a phosphate group
   (b) Nucleotides polymerise through phosphate ester linkages to form nucleic acids (see Figure 5.2C).
   (c) replication of DNA (Figure 5.5A), transcription of mRNA from DNA (Figure 5.3B) and polymerisation to form tRNA (Figure 5.3A)
6. Semi-conservative. After one replication, all the ^{15}N-DNA is paired with ^{14}N-DNA, supporting Figure 5.5A. All the double helices contain one strand of ^{15}N-DNA and one strand of ^{14}N-DNA. After two replications, half the DNA

molecules contain one strand of ^{15}N-DNA bonded to one strand of ^{14}N-DNA and the other half contain a double strand of ^{14}N-DNA. After three replications, one quarter of the DNA molecules contain one strand of ^{15}N-DNA.

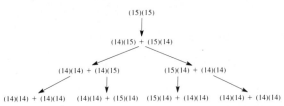

Checkpoint 5.6
1. (a) Daughter cells need to have the same genes as the parent cells.
   (b) a mutation – when a wrong nucleotide bonds to the template
2. For (a) and (b) see §5.6
   (c) See §5.5, including Figure 5.5A. If a wrong nucleotide bonds to the template, a mutation arises.

Questions on Chapter 5
1. See §5.2
2. (a) AUGCCUAAGUAC
   (b) See Figure 5.4A
3. (a) See §5.2, including Figures 5.2B, C, D and E
   (b) A triplet of bases on DNA codes for a certain amino acid. The triplet is transcribed into a codon in mRNA (Figure 5.3B). A codon in mRNA bonds to an anticodon in a molecule of tRNA. The molecule of tRNA bonds to the amino acid for which the anticodon codes (Figure 5.4A). In short, DNA triplet → mRNA codon → tRNA anticodon → amino acid
4. Altering the genes of an organism by transplanting into it a section of DNA from another organism; see §5.8 and for examples also see §5.8
5. mRNA, tRNA, rRNA; see §5.3

6. See § 5.2, including Figure 5.2C
7. DNA has deoxyribose where RNA has ribose; DNA molecules are double-stranded, but RNA has single-stranded molecules (with base-pairing in some regions); DNA contains adenine, thymine, cytosine and guanine, whereas RNA contains adenine, uracil, cytosine and guanine.
8. A triplet of nucleotides on DNA codes for a certain amino acid. The length of DNA that codes for a whole protein is a gene. By controlling protein synthesis (see Figure 5.4A). DNA decides which proteins will be synthesised in an organism, including the enzymes that control the metabolism of the organism.
9. See § 5.4, including Figure 5.4A
10. (*a*) double helix, base pairing; see § 5.5
    (*b*) In mitosis daughter cells must have exactly the same DNA as parent cells.
    (*c*) See § 5.5 and Figure 5.5D
11. (*a*) **A** monosaccharide, **B** amino acid,
        C deoxyribonucleotide
    (*b*) **A** polysaccharide, **B** peptide, polypeptide, protein,
        C deoxyribonucleic acid, DNA
    (*c*) **A** a source of energy, **B** for construction and repair,
        C carrying genetic information
    (*d*)

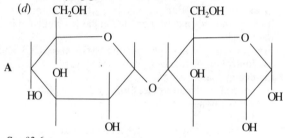

See § 3.6

B $H_2NCHCONHCHCO_2H$

   $HO_2C(CH_2)_2$ $(CH_2)_2CO_2H$

See § 2.2

C

See Figure 5.1A

CHAPTER 6: VITAMINS

Questions on Chapter 6
1. vitamins B_{12}; he could take a yeast extract, e.g. Marmite.
2. riboflavin

3. vitamin D and possible vitamin A
4. 30 mg
5. (*a*) See § 2.4   (*b*) See § 3.4   (*c*) See § 4.2   (*d*) See § 6.2

CHAPTER 7: MINERALS

Questions on Chapter 7
1. See § 7.2
2. (*a*) iron   (*b*) § 7.3
3. See § 7.2
4. See § 7.5
5. (*a*) See § 7.6

(*b*) The family members have the same diet, which lacks iodine.
6. (*a*) In the fluoridated compound there is less —OH to be attacked by mouth acids.
   (*b*) See § 7.7

CHAPTER 8: WATER

Questions on Chapter 8
1. See *ALC*, § 9.5
2. See § 8.3
3. See § 8.3

4. (*a*) Monosaccharides and disaccharides have —OH groups which can form hydrogen bonds with water.
   (*b*) Lipids do not contain —OH groups or polar groups.
5. See channel proteins; § 4.12 and Figure 4.12B

CHAPTER 9: METABOLIC PATHWAYS

Checkpoint 9.2
1. (*a*) low
   (*b*) The cell will respire glucose with the formation of ATP.
   (*c*) by enzymes which are inhibited by ATP (§ 9.2)
2. CH_3COCO_2H, 2
3. The energy released by direct oxidation of glucose would overheat the cells. Cells require a continuous supply of energy (see § 9.1).

Checkpoint 9.4
1. (*a*) acetyl coenzyme A and citrate ion   (*b*) 4   (*c*) See § 9.3
2. (*a*) glycolysis and the tricarboxylic acid cycle
   (*b*) Pyruvate ion + Co A + NAD^+ →
       Acetyl Co A + CO_2 + NADH + H^+
   (*c*) coenzyme A, which combines with ethanoate (acetate) ion and NAD^+ which combines with hydrogen
3. (*a*) the link reaction; see § 9.3
   (*b*) transfer of hydrogen and electrons to oxygen; see § 9.4

Checkpoint 9.6
1. (*a*) See § 9.2
   (*b*) conversion into ADP + P_i; see § 9.2
   (*c*) The conversion of ADP + P_i into ATP is endothermic, and the conversion of ATP into ADP + P_i is exothermic; thus ATP is able to act as an energy 'bank'; see § 9.5
2. (*a*) (i) 38 mol   (ii) 2 mol
   (*b*) In aerobic metabolism, glycolysis is followed by the TCA cycle, which produces more ATP.
   (*c*) The oxidation of lactic acid to carbon dioxide and water yields additional energy.
3. (*a*) the conversion of glucose into ethanol + carbon dioxide; catalysed by enzymes in yeast
   (*b*) lactic acid
   (*c*) Yeast is killed by this concentration of ethanol.

Checkpoint 9.8
1. For (*a*) and (*b*) see § 9.8   (*c*) See Table 9.8
2. See § 9.8
3. See § 9.8.4

Questions on Chapter 9
1. (*a*) Lactate is formed when oxygen is in short supply and anaerobic respiration occurs.
   (*b*) An oxygen debt has been built up.
   (*c*) Glycogen can be converted into glucose and metabolised when needed.
   (*d*) The supply of glucose in blood will soon be used up.
   (*e*) E.g. it increases the capacity of the lungs and improves the supply of oxygen.
2. (*a*) See Figure 9.3B
   (*b*) See § 9.6
   (*c*) See § 9.3, § 9.5
   (*d*) See § 9.4
3. Being a polypeptide, it is hydrolysed by proteolytic enzymes in the gut.
4. (*a*) glycolysis
   (*b*) $C_6H_{12}O_6$ → $2CH_3COCO_2^-$ + $6H^+$
   (*c*) in the cytosol
   (*d*) aerobic
   (*e*) tricarboxylic acid cycle, § 9.3; electron transport chain, § 9.4
   (*f*) $2CH_3COCO_2^-$ + $6H^+$ + $6O_2$ → $6CO_2$ + $6H_2O$
5. Glycolysis is occurring. Jane is short of oxygen and has built up an oxygen debt. Lactate ion has accumulated in her muscles.
6. For interconversion of glucose and glycogen see § 3.5, and for control of the rate of glycolysis see § 9.2.
7. (*a*) energy 'bank'; see § 9.5
   (*b*) control of glycolysis (see § 9.2), oxidative phosphorylation (see § 9.4)
   (*c*) DNA
8. (*a*) It changes from orange through green to blue.
   (*b*) oxidation of ethanol
   (*c*) Weigh before and after fermentation.
   (*d*) $C_6H_{12}O_6$ → $2C_2H_5OH$ + $2CO_2$
   (*e*) pass air instead of nitrogen.
   (*f*) No change because the products would be carbon dioxide and water.

PART 2: FOOD SCIENCE

CHAPTER 10: FOOD QUALITY

Questions on Chapter 10
1. (*a*) See § 10.1
   (*b*)

2. (*a*) plant cell wall (§ 10.2)
   (*b*) cellulose and pectic substances (§ 10.2)
   (*c*) skeletal muscle (§ 10.3)
3. See § 10.3
4. See § 10.3 and Figure 10.3E
5. See end of § 10.3

CHAPTER 11: FOOD PROCESSING

Checkpoint 11.1
1. (a) See § 11.1
   (b) Na^+ cannot bond to two —CO_2^- groups.
2. For (a) and (b) (i) see § 10.6.
   (b) (ii) See § 11.1
   (c) Taste and nutrients are conserved in microwave cooking.

Checkpoint 11.3
1. (a) See Figure 11.2A
   (b) See § 11.2
   (c) and (d) See § 11.3
2. For (a), (b), (c) and (d) See § 11.2
   (e) for yeast to ferment
   (f) See § 11.3
3. $25.4 \, cm^3$
4. $30.6 \, cm^3$
5. NH_4HCO_3 dissociates rapidly to form
   $NH_3 + CO_2 + H_2O(g)$. A very hot oven makes the water vaporise, and the vapour leavens the mixture.
6. (i) Self-raising flour contains sodium hydrogencarbonate. Add acid and test for CO_2.
   (ii) Hard wheat flour contains more protein than soft wheat flour. Assess the size of the gluten ball as described in § 11.2
7. For (a), (b), (c) see § 11.3; for (d) and (e) see § 11.2

Questions on Chapter 11
1. (a) See Figure 11.2A   (b) See § 11.2   (c) See § 11.3
   (d) See § 11.3
2. (a) for carbon dioxide production through fermentation by yeast
   (b) See § 11.2 on Maillard browning
   (c) See § 11.3
3. (a) oxidation of pigments in flour to give a whiter crumb and oxidation of —SH groups to —SO_3H; increasing the elasticity of the dough (§ 11.2)
   (b) See tests for bromate (v), ascorbic acid and iron in § 11.2
4. See § 11.2
5. (a) Hard wheat (§ 11.2) contains more protein than soft wheat. Hard wheat gives strong flour, which makes an elastic dough, suitable for bread-making. Soft wheat gives weak flour, used for cakes and pastry.
   (b) See § 11.3
6. (a) Meat from a slaughtered animal contains more lactic acid (§ 11.3).
   (b) papain, a proteolytic enzyme
   (c) (i) oxymyoglobin   (ii) nitrosomyoglobin

CHAPTER 12: FOOD PRESERVATION

Checkpoint 12.4
1. to raise the boiling temperature of water above 100 °C in order to kill *Clostridium* etc.
2. slows down the growth of micro-organisms but does not kill spores.
3. Osmosis draws water from the food into the sugar solution, leaving less water available for micro-organisms.
4. Lactic acid inhibits the growth of micro-organisms.
5. Both remove water which is necessary for the growth of micro-organisms.
6. Food retains its shape and appearance.
7. See § 12.4 'Milk'.
8. (a) to remove water by osmosis
   (b) inhibits the growth of micro-organisms
9. (a) Raising the temperature above 121 °C for 10 minutes kills all micro-organisms, including spores of *Clostridium*.
   (b) Refrigeration slows down the growth of micro-organisms.
   (c) Freeze-drying removes most of the water needed by micro-organisms.
10. Canning at 121 °C for 10 minutes is entirely safe: it kills all micro-organisms. Food can be kept for a very long time without the expense of refrigeration.
11. (a) See § 12.4
    (b) botulinum cook, 121 °C for 3 minutes
12. (a) See § 12.3
    (b) e.g. cheese-making, yoghurt-making; see § 12.3
13. Immerse in salt solution or sugar solution or acid solution; dry; exclude air; blanch and freeze; cook; for reasons see § 12.4
14. See Figure 12.3A
15. See § 12.4.
    (a) freezing > canning > dehydration
    (b) canning > dehydration > freezing
    (c) freezing > canning > dehydration
16. fermentation – lactic; proteolysis – amino – ammonia – hydrogen sulphide; lipolysis – fatty – carbonyl

Questions on Chapter 12
1. (a) e.g. *Salmonella, Clostridium, Staphylococcus*
   (b) e.g. moulds and fungi
   (c) e.g. cheese-making; see § 12.3
   (d) See Figure 12.3A
2. (a) Browning would not occur at the surface only if it were due to micro-organisms: it would occur inside the fruits and vegetables.
   (b) Exclude air by immersion in boiled-out water.
   (c) Increases with temperature and then comes to a stop above a certain temperature (at which enzymes are denatured). It is arrested by enzyme-inhibitors.
3. controversial; see § 12.4
4. Dried milk at 535 p/kg is the most expensive because the customer is not paying for the water content of whole milk. Condensed milk at 198 p/kg comes next, followed by evaporated milk at 124 p/kg, again in order of reduced water content. Sterilised milk at 60.0 p/kg and pasteurised milk at 37.9 p/kg contain more water and are cheaper. Sterilisation is costlier because it involves longer heat treatment at a higher temperature.
5. (a) See § 11.2–3 for oil content, cellulose content and vitamin content of bran and germ which are removed in making white flour.
   (b) Amylose and amylopectin in wheat are hydrolysed to maltose; see § 11.3
   (c) Moulds need oxygen (§ 12.3). Aerobic bacteria need oxygen. An atmosphere of carbon dioxide limits the growth of moulds and aerobic bacteria.

CHAPTER 13: ADDITIVES

Questions on Chapter 13

1. (*a*) to 'cure' the meat and inhibit the growth of bacteria. The consumer and the vendor benefit.
   (*b*) The vendor of the ham benefits because he can inject water into it.
   (*c*) Sodium nitrite makes the ham safe to eat; sodium heptaoxophosphate(v) makes a bigger profit for the vendor.
2. (*a*) (i) to prevent the growth of micro-organisms
        (ii) to prevent the autoxidation of lipids
        (iii) to impart an appetising yellow colour
        (iv) to enable water to be retained in the cheese spread.
   (*b*) There must be a substantial water content.
3. (*a*) See § 13.2
   (*b*) See § 13.4, § 4.14
   (*c*) See § 4.8; raises the melting temperature
   (*d*) Make food look more attractive. Some people are allergic to tartrazine; see § 13.5

CHAPTER 14: LEGISLATION

Questions on Chapter 14

1. See § 14.3
2. a description of the food, the name of the manufacturer, the sell-by date and the net weight
4. (*a*) See §§ 13.1–4
   (*b*) See § 13.5, § 14.1
5. See § 14.2

CHAPTER 15: HUNGER

Questions on Chapter 15

1. (*a*) fertilisers; see *ALC*, § 22.2.1, § 22.7.3
   (*b*) insecticides and herbicides; see *ALC*, § 20.15, § 32.11
2. Short-term aid and long-term aid are both needed; see § 15.3
3. In this particular case, the Ethiopian government was using income to pay for arms to fight a civil war so a firm refusing to buy the meat would not have been making the plight of the Ethiopian people worse.
4. Different situations obtain in different countries.
5. There is no one correct answer!
6. All three are wise moves. In the short·term (*b*) is wise; in the long term (*a*) and (*c*) will pay.
7. Various reasons have been put forward in Chapter 15.
8. The humanitarian answer is to feed the people, but it is a matter for discussion how to do this and also develop the country's economy.
9. Capital to instal the fermenter, clean water, electricity and trained personnel to run it. It could be useful where land is not available for agriculture and where the climate is harsh.

Index